# Stars Over the Red Cedar

## Research, Education, and Outreach at the Michigan State University Campus Observatories

### Horace A. Smith

Chapbook Press

Schuler Books
2660 28th Street SE
Grand Rapids, MI 49512
(616) 942-7330
www.schulerbooks.com

Stars Over the Red Cedar: Research, Education and Outreach at the Michigan State University Campus Observatories

ISBN 13: 9781948237406

Library of Congress Control Number: 2019918981

Front cover: The 1970 observatory after the ice storm of December, 2013.

Back cover: The 1880 campus observatory photographed circa 1888. Courtesy of MSU Archives and Historical Collections.

Title Page: Northern lights over the observatory during the severe "Halloween" geomagnetic storm of 2003.

**Printed in the United States by Chapbook Press.**

## Table of Contents

## Acknowledgments

Thanks are owed to many who provided their memories, thoughts, and photographs concerning the Michigan State University campus observatories and related matters. Among those deserving acknowledgment are Robert Miller, Mary Anderson, Nancy Silbermann, John Briggs, John French, Wayne Osborn, Pamela Gay, Trudy Bell, Karen Kinemuchi, Mary Gowans, David DeVorkin, Laura Chomiuk, David Batch, Susan Simkin, Cameron Dean, Rodney Dean, Marshall Dean, and Deborah Benedict.

I especially thank the staff of the Michigan State University Archives and Historical Collections, both for their help in tracking down historical items and for permission to reproduce photographs in their collection. The Smithsonian Astrophysical Observatory Archives likewise provided important information on Lansing's Moonwatch station.

Not everyone connected with the work of the observatory is mentioned by name in the following chapters. That is particularly true for the numerous students who were engaged in outreach and research there, only a few of whom are explicitly acknowledged. Many of their names can, however, be found in the Appendix, where they appear among the authors of papers that grew from their endeavors at the telescope.

# I. Introduction

The 24-inch[1] (61-cm) Boller and Chivens telescope, main instrument of the Michigan State University campus observatory, saw first light in 1970. In 2020, fifty years will have elapsed since its installation. During those years the observatory has been employed for astronomical research, education, and public outreach. It has also survived periods of financial hardship that strained university resources and growing light pollution at its south campus location.

It was the approach of the 50th anniversary of the 1970 observatory that inspired the writing of this history. Nevertheless, the narrative is not limited to the story of that observatory alone. It also examines its nineteenth century predecessor and tells of satellite-observing by citizen-scientists at the dawn of the space-age. This account is, however, by no means a comprehensive history of astronomy, or even just observational astronomy, at Michigan State University. For example, Michigan State University's participation in the SOAR telescope project in Chile appears only as an aside. The

---

[1] I will often use English units to describe the observatory telescopes as such units were commonly used by those involved. Thus, the 24-inch rather than the 61-cm telescope.

book also makes no attempt to convey the story of Abrams Planetarium, a center of education and outreach on campus deserving of its own history.

Some motivation may be deemed necessary for a history of such modest observatories as have adorned the Michigan State University campus. After all, a 24-inch telescope by no means counts as a major astronomical instrument today, nor could it have been called such even when it was built. It is, for example, much smaller than the 4.1-m SOAR telescope and SOAR itself is dwarfed by 8-m and larger telescopes now in operation. However, many undergraduate students, some of them future astronomers, gained their first acquaintance with research at the campus observatory, and many members of the public obtained their first telescopic views of astronomical objects through the 24-inch reflector. Such experiences with small telescopes are by no means unique to Michigan State University, but they are too seldom remembered in print.

Moreover, as the 20th century drew to a close and the 21st century began, ground-based telescopes of less than 1-m aperture proved to be surprisingly durable research instruments (see, for example, Percy 1986 and Stephens et al. 2016). The changing niches open for research with small telescopes, and the changing instrumentation employed on them, are worthy of consideration.

While this book describes how astronomical observing on campus has changed during the past century and a half, it emphasizes the fifty-year history of the 1970 observatory. Appendix I provides background on the research programs of that observatory, but details of its various research projects are beyond the scope of this volume. Those interested in delving more deeply into why faculty and students lost sleep at the observatory are directed to the papers listed in Appendix II, most of which are freely available through the SAO/NASA Astrophysics Data System.

Finally, this history is valuable as an acknowledgement of the people involved, directly and indirectly, in astronomical observing from the campus over the course of more than 140 years.

*The 1970 campus observatory after the ice storm of December, 2013. Author's photograph.*

## II. 1880: The First Campus Observatory

As the present observatory celebrates its 50th year, 140 years will have elapsed since an earlier, long vanished, observatory first graced the campus. The observatory that today sits between the Forest Akers Golf Course and the Pavilion for Agriculture and Livestock Education, south of the Red Cedar River, was not the first associated with the college that would eventually become Michigan State University. It had a predecessor, built in 1880, at a time when many small observatories were being built in the United States.

The Agricultural College of the State of Michigan, as it was first known, was a pioneering land grant institution. It was founded in 1855, although classes did not begin until two years later. In 1880, the college, by then named simply the State Agricultural College, was still relatively new and very much a junior institution compared to the University of Michigan, which assumed that name in 1821 and moved to its present home in Ann Arbor sixteen years later.

The construction of observatories in the United States had a slow beginning (Musto 1967), but the pace of building picked up in the second half of the nineteenth century (Bell 2002, Cameron 2010). As the 1870s ended, the University of Michigan already had an observatory, the Detroit Observatory, which was built in 1854 and which housed a 12 $5/8$-inch Fitz refracting telescope and a 6-inch meridian circle telescope (Whitesell 1998). While those instruments were designed from the beginning with research in mind, at that time the future Michigan State University had no plans for astronomical research and, in any case, no observatory with which to conduct any. Nonetheless, an interest in astronomy was not absent.

*Professor Rolla C. Carpenter from the Class Album of 1885. Michigan State University Archives and Historical Collections.*

Astronomy and surveying were added early to the Agricultural College's curriculum.[2] Professor Rolla C. Carpenter (1852 – 1919), who taught mathematics and civil engineering, also taught astronomy, and it was thanks to him that the college built its first observatory.

Carpenter was appointed to the college faculty in 1875, and soon was meeting students atop the college's boarding hall[3], where they observed celestial objects with a small portable telescope (Carpenter 1876a). For the next few years, his reports on the activities of the Department of Engineering and Mathematics signaled an increasing focus on astronomy.

In his report for 1876, Carpenter wrote that "Opportunity was afforded the [astronomy] class for observing Jupiter, Saturn, and the moon, with the telescope now on deposit at the College" (Carpenter 1876b). Two years later, Carpenter would write that "Astronomy was taught six weeks during the third term of 1877 to the junior class of that year, consisting of 28 members. The class exhibited a great deal of enthusiasm both in their class recitations and in their evening work of observing planets and constellations...Our means of

---

[2] Suelter (2008) discusses the evolution of astronomy courses and teaching during the early years of the university.

[3] Probably the so-called new boarding hall that opened in 1870 and which was later named Williams Hall is meant, although "boarding hall" could also mean the old boarding hall, Saints Rest dormitory, which burned at the end of 1876 (see Beal 1915).

illustration in this study is very meager and consist entirely of borrowed apparatus, viz: a small telescope and a celestial globe. A good telescope, properly mounted, is needed very much" (Carpenter 1878). Carpenter's enthusiasm for astronomy seems to have at least matched that of his students.

Professor Carpenter's plea did not go unheeded. With the approval of President Theophilus Abbot, Carpenter purchased an "astronomical telescope with a 5-inch object glass of Alvan Clark and Sons, $450" (Carpenter 1879). The sum was not an inconsequential one in 1879, being about a third the annual salary of a faculty member. Although an aperture of 5-inches is specified, later reports will, as we see below, sometimes refer to the slightly greater aperture of 5½-inches. Because, as we shall also see, this telescope is still in the hands of the university, I was able to measure the actual diameter of the objective, finding a clear aperture of 5 inches to be correct.[4]

Such a fine telescope called for a permanent site for its use. The State Board of Agriculture (essentially the college Board of Trustees) on June 9, 1880, "resolved

---

[4] The full diameter of the lens is partly hidden by the lens cell and is slightly larger than 5-inches. That may be the origin of the 5 ½-inch number. Antique telescope expert John Briggs suggested (based upon photographs) that the front part of the objective cell is not now the original. He wondered whether the Clark lens could have been replaced in the 20th century but I have found no evidence of such a change.

that Prof. Carpenter be authorized to construct a small brisk [brick?] house in which to mount the telescope from the appropriation to the Mechanical De[pt.] [and the] Mathematical Dep't. the Structure to be located by Pres. Abbot." [5]

Professor Carpenter wasted no time and, before the year was out, he was able to describe what the Board had gotten for their money: "The building is located on the hill northeast of President Abbot's house.   It is circular in plan with an external diameter of 16 feet, and an internal diameter of 14 feet 8 inches. The walls are of brick, nine feet high, [and] bear on their top a wooden plate on which is fastened an iron track, on which the roof can be made to revolve.   The roof has 14 hips and terminates in a pendant at the center.   It is mounted on wheels and has one trap door which can be turned so that an observer can see in any possible direction from the inside. The roof can be rotated either by a lever from the outside, or by a windlass and pulley from the inside. The building is lighted by two windows and is ventilated by doors placed in the frieze.

"The whole building is on a concrete foundation and is built in the most thorough manner. Its extreme height is 16 feet. The telescope is mounted on [a] brick

---

[5] Notes from early meetings of the State Board have been posted on the website of the Michigan State University Archives and Historical Collections.

pedestal five feet in diameter, which is entirely independent of the building. The inside of the building is unfinished; it will do for present use, but it should be finished as soon as money can be spared." The "pedestal" must refer to a firm pier on which the telescope could be set rather than a complete mounting, for Carpenter continues: "The telescope which now stands on a tripod should be mounted more substantially and provided with vertical and horizontal circles for reaching angles. The cost of the building in its present condition was about $125." The low cost may have been in part because the "carpenter work was done almost entirely by students under my supervision" (Carpenter 1880).

Besides being near President Abbot's house, the observatory was also close to Carpenter's Faculty Row Number 2 dwelling. Its location was a little south of the still existing Wills House, built in 1927 for the Weather Bureau, and was, like almost all of the campus at the time, north of the Red Cedar River.[6] Light pollution was not a problem one had to contend with at that time when siting an observatory. The skies, when clear and moonless, would be dark.

---

[6] In the Smithsonian *Report on Astronomical Observatories for 1886* the observatory's longitude is given, in time units, as $29^m44^s$ west of Washington, DC, and its latitude as 42°43'54" north. A check of the approximate observatory site using Google Earth (WGS84 geodetic datum) gives latitude 42°44'03" north and longitude $84°29^m13^s$ west.

Perhaps the trustees were pleased with Carpenter's frugality in the construction of the observatory. On October 10, 1881, the Board further authorized Carpenter to spend an amount not to exceed $400.00 in the purchase of a proper mounting for the telescope. In his departmental report for 1880-1882, Carpenter would note that the telescope (now referred to as a 5½-inch instrument rather than a 5-inch) was "equatorially mounted with clock work and declination circle reading to 30 seconds of arc and hour circle reading to single seconds of time" (Carpenter 1882). Photographs, dated 1888 and 1909, show the telescope with just such an equatorial mounting and weight driven clock-drive, firmly seated on a central pier.

The equatorial mounting is not a usual one for an Alvan Clark and Sons telescope of that time. John W. Briggs, an expert on historical telescopes, led me to the manufacturer of the mounting, the British firm of Richard and Joseph Beck.[7] An illustration in their 1882 brochure depicts a telescope with a mounting very similar to that in the photographs showing the 5-inch telescope. One might wonder why the Beck firm, based

---

[7] Trudy Bell pointed out to me that the 1883-1884 college catalogue (p. 27) explicitly states that the telescope was mounted by "Beck of London", confirming the Beck connection. Furthermore, the *College Speculum* for August 1, 1882 reported that the telescope mounting had recently arrived from London.

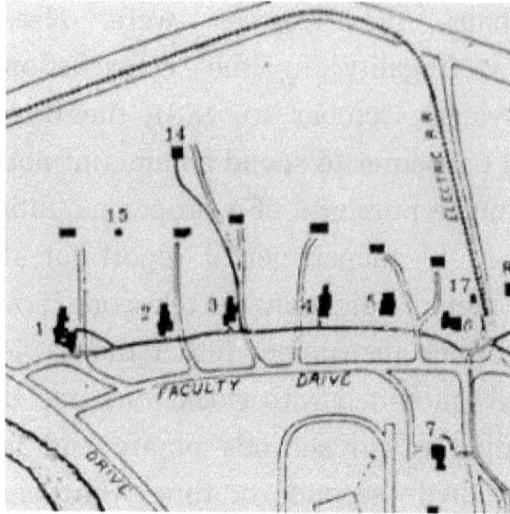

*A portion of an 1899 map of the college campus. Building 15 is the first observatory. Professor Carpenter lived in the Faculty Row house labeled number 2. Michigan Avenue is the street directly above the observatory. North at top. East at right. Michigan State University Archives and Historical Collections.*

in London but with a branch in Philadelphia, was selected as the source for the telescope mounting. The answer may lie with a specialty of the firm, microscopes, which the Agricultural College was also buying around 1880. It is plausible, but not established, that the college investigated the purchase of Beck microscopes, incidentally leading to an awareness of the R. and J. Beck line of telescope products. Carpenter's comments on the precision of the declination and hour angle circles duplicate those in the Beck brochure.

The Board on June 3, 1889, would make one more provision for the observatory. It allowed Professor Carpenter to spend $50 on a micrometer for the telescope. The purpose of the micrometer is not specified, but it may have been a filar micrometer for accurate positional measurements. The Fauth and Company 1883 catalog offered an "Eye-piece micrometer (filar)" for $50.00.[8]

The *College Speculum*, an early college newspaper, for August 1, 1881, appreciated Professor Carpenter's efforts: "Prof. R. C. Carpenter deserves credit for getting a telescope and accessories in good working trim at the College. The instrument is a fine one, although rather small – 5½ inches. It is manufactured by the celebrated Alvan Clark and Son[s]. The observatory is located just northwest of the Professor's residence, is of brick, with movable roof. The telescope will soon be mounted on clock work, the necessary means being now at hand[9]." The wording is closely

---

[8] https://archive.org/details/catalogueofastro00faut (downloaded 10/24/2017)

[9] Interestingly, this comment in the August 1 *Speculum* is dated before the Board's October 10 authorization for the mounting. However, the February 15, 1881, Board minutes did mention "Telescope Mountings $100" under "Estimates...for Extraordinary Expenses for the years 1881 and 1882".

similar to that used in William J. Beal's history of the college (Beal 1915), perhaps not surprisingly since Beal was the *Speculum*'s editor for science.

Musto (1967), in his survey of the American observatory movement between 1800 and 1850, identified three stages in development. The first included several failed attempts to establish a permanent research observatory in the United States. The second stage proved more promising. Musto stated: "During the next decade, until 1840, a number of small permanent observatories were built under the auspices of several colleges and the Federal government. Each was, by and large, the work of a single person given modest financial support by his employer." The third stage, arriving in the 1840s, saw the successful establishment of more substantial permanent observatories, supported by colleges, communities, wealthy individuals, and the Federal government.

The 1880 observatory was built subsequent to 1850, the final year of Musto's three stages. Although it bears a resemblance to observatories established during Musto's second stage of observatory development, by this later time small observatories were springing up in many locations. By the 1880s, such facilities were no longer rarities in the United States, or in Michigan for that matter.

*The 1880 campus observatory from the photograph in Beal's (1915) history of what was then called Michigan Agricultural College. The photograph is undated, but a time in the early 1900s is plausible given the date of Beal's book.*

In Michigan, other small college observatories were put up both before and after the one at the Agricultural College. In 1878 Michigan State Normal College (now Eastern Michigan State University) in Ypsilanti mounted a four-inch Clark refractor in an observatory atop the college's Pierce Hall, while Albion College in 1883-1884 built an observatory housing an eight-inch Clark refractor. Like the Agricultural College observatory, those observatories were built for the purpose of education rather than research. Refracting telescopes rather than reflectors dominated, with telescopes by Alvan Clark and Sons being the favorites of the day.[10]

Carpenter would remain at the State Agricultural College until he moved to Cornell University in 1890. The observatory he built was used for the instruction of students, but, with one exception mentioned below, it does not appear to have been employed in any sort of research.

The 1880 campus observatory was apparently used for daytime as well as nighttime observing and here we come to the only evidence I have found of the observatory being used for a scientific purpose. For several months in 1885, the *Monthly Weather Review* reported sunspot counts made by Professor Carpenter of the Michigan State Agricultural College. The reports do not actually specify that the observatory telescope was used for these observations, but such is strongly suggested by a report in the *Lake County* [Michigan] *Star* for June 25, 1885, in which it was stated that "Observations taken recently at the Agricultural College observatory revealed sixty spots in the sun's surface. Two of them were over 20,000 miles in diameter."

---

[10] In 1880, the Battle Creek, Michigan, high school reported having a 4-inch refracting telescope, but whether it was within an observatory structure that early is not clear to me. (Report on Astronomical Observatories, Smithsonian Institution, 1886.)

One might be forgiven for thinking that the Professor Carpenter who made the solar observations was Rolla C. Carpenter. When a fuller name is given in the *Monthly Weather Review*, it is instead that of Professor L. G. Carpenter. That, presumably, is Rolla's brother, Louis George Carpenter, who taught mathematics at the Agricultural College from 1883 until 1888, before leaving to take a position at the Colorado Agricultural College.

When Rolla Carpenter left for Cornell in 1890, the greatest champion of the observatory may have departed, but even before he left, increasingly heavy demands on his time had been drawing his attention elsewhere. After Carpenter's departure, Herman K. Vedder assumed the professorship of Civil Engineering, and with it responsibility for the observatory. His direction of the observatory would continue into the next century. Under Carpenter and Vedder, the first MSU observatory saw steady student use with little fanfare. It would meet a sorry end in the first quarter of the twentieth century.

*The interior of the first observatory showing the telescope's equatorial mounting, weight-driven clock drive, and brass tube. The larger object on the table is a Bailey's Astral Lantern. It illuminated maps depicting the night sky to help students identify stars and constellations. Photograph dated 1909. Michigan State University Archives and Historical Collections. Resource identifier A000122.jpg.*

*Photograph of the observatory c. 1888. I believe that Rolla Carpenter may be standing near the center with his hand to his head. Could this be a class of astronomy students? And what do the summer hats (straw boaters?) worn by many signify? Michigan State University Archives and Historical Collections. Collection number UA 10.3.265. The original photograph has been slightly cropped to better fit the page.*

*Note the wheels and circular track upon which the wooden dome of the 1880 observatory rotated. An extended dew cap, accessory, or cover is present that is not seen in the other photographs. Photograph by L. G. Carpenter, dated 1888. Michigan State University Archives and Historical Collections.*

The 1909 photograph (accepting as correct the date on its back) shows the observatory in good condition at the close of the first decade of the 20th century. However, a 1913 map shows that a driveway or road had been constructed from Michigan Avenue that swung closely by the observatory. That ease of access may have been a bad thing. A letter written by Dean of Engineering G. W. Bissell to President J. L. Snyder, dated June 7, 1915, conveyed the news that the observatory had been burglarized, and that considerable equipment had been taken or destroyed.

More about this sorry event was relayed in that year's annual report for the Department of Civil Engineering: "As I informed you at the time, about at the close of the spring term some person or persons maliciously inclined entered our little astronomical observatory and despoiled its contents of practically everything movable, besides wantonly destroying other parts of the equipment that could not be carried away. About the only thing left uninjured is the excellent 5½ inch Clark objective of the equatorial telescope. I wish to recommend that the equatorial be provided with a new mounting and a driving clock. It goes without saying that the observatory should be made secure if possible. There is evidence to indicate that those who committed the above depredations entered by the doorway; hence they must have been in possession of a key or were able to successfully pick the lock" (Vedder 1915). It sounds as though we are lucky that not all was lost!

Dean Bissell suggested that "a detective be hired to find the guilty party", but I have come across no evidence that the miscreant or miscreants were ever apprehended.

The burglary and vandalism of 1915 appear to have spelled the end of the first observatory. The disastrous fire which destroyed the engineering building in March, 1916, soon brought far more urgent needs to the engineering program than making good the old observatory. Moreover, the following year saw the entry of the United States into World War I, with new demands upon the university. It was not a good time for refurbishing the almost forty-year-old observatory.

In any case, the observatory was apparently demolished not long thereafter, although I have not found a precise date for its removal. The 1915 letter and report describing the burglary certainly indicate that it existed until that time. The catalog description of the astronomy course for engineers for 1920-21 no longer mentioned two hours of observatory work a week, as had prior catalogs, but two hours of field work. The observatory was still listed on the building inventory in the college's annual report for the year ending June 30, 1923, but valued at only $100. It is gone from the 1924 building inventory and was not included in a 1926 map[11] of the campus, so the early 1920s are a plausible time for its end.

It was fortunate that the Clark 5-inch telescope objective and the tube in which it was mounted were not carried off by burglars and survived the loss of the observatory. In the early 1970s, then graduate student Robert D. Miller and Astronomy Professor Thomas R.

---

[11] *City of East Lansing Use Districts.* Drawn by Hubbell Hart Gering & Roth and published by the City of East Lansing in 1926.

Stoeckley re-discovered the telescope in a wooden box in a storage room on the third floor of the building which then housed the Physics and Astronomy departments.[12] The telescope's clock drive, finder scope, Beck mounting, and accessories were not found with the telescope and are presumably among the items lost in 1915. As of this writing, the telescope is on exhibition at Abrams Planetarium.

The telescope now has an equatorial mounting atop a short tripod. John Briggs suggested that parts of the present mounting derive, not from the missing Beck mounting, but instead from a mounting supplied with portable Alvan Clark and Sons telescopes. This opinion finds support in photographs of such mountings in the book on Alvan Clark and Sons telescopes by Warner and Ariail (1995, p. 225). Might those parts have come from the original tripod mounting for the 1879 telescope?

---

[12] I have not been able to discover whether the telescope was ever used for classes after the demise of the 1880 observatory. It was, however, apparently at one time sneaked from Abrams Planetarium for a star party.

*The telescope of the first observatory as it looked in 2018. Courtesy Robert Miller.*

*The inscription on the telescope reveals its maker and when it was made. Not shown is a second inscription, "Cambridgeport, Mass.," telling where it was made. Courtesy Robert Miller.*

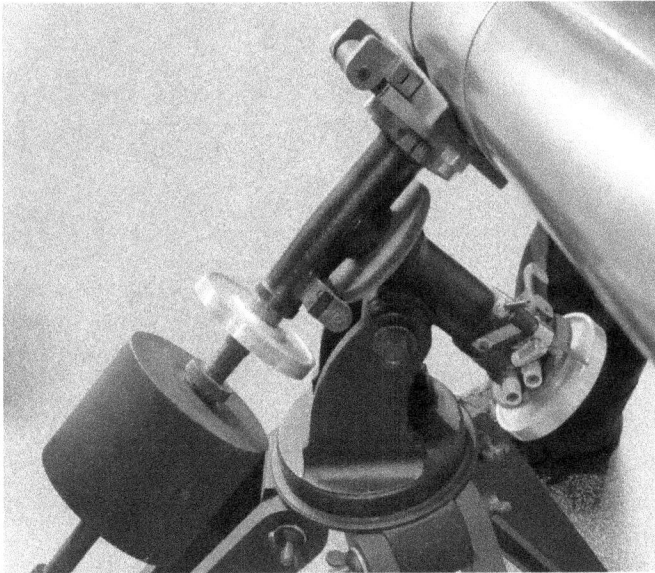

*The equatorial mounting now supporting the 5-inch telescope. Author's photograph.*

This rather sparse information about the first campus observatory is all that I have been able to find. We have only slight hints as to how the observatory was actually used. In his departmental report for 1884, Professor Carpenter remarked that "Astronomy was taught during the summer term to such Seniors as wished to take it. It was chosen by a large proportion of the class. Newcomb's *Popular Astronomy* was used as a text book, but the time of the class was largely devoted to practical work in our observatory, and in the study of constellation figures" (Carpenter 1884).

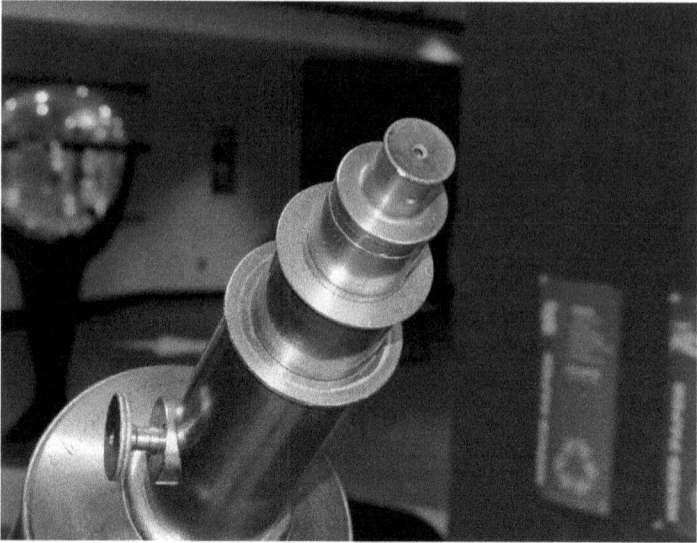

*Tailpiece of the 5-inch telescope in 2018, with its sole remaining antique eyepiece. Author's photograph.*

Alas, I have come across no firsthand or even second-hand account from anyone about what it was actually like to look through the Clark telescope in its heyday. Surely some of the sights shown to students were the same as excite small-telescope observers today: the craters of the Moon, the Great Nebula in Orion, the rings of Saturn.

In addition, there were transient spectacles that might have livened observing near the time the first observatory was built and in the mind's eye one can imagine people eagerly gathered around the eyepiece viewing them. Early in the summer of 1881 the Great Comet of that year (C/1881 K1) was visible, its tail extending far beyond the field of view of the 5-inch telescope but its coma making a fine telescopic object. If they missed that comet, perhaps members of the college community were able to see instead the Great September Comet of 1882 (C/1882 R1) – the early 1880s were a good time for comet watchers.

Or we might imagine Professor Carpenter and students scrutinizing the image of Jupiter to see the Great Red Spot, large and prominent and strongly red as the 1870s ended and the 1880s began (Rogers 1995, Hockey 1999). Famed astronomer Edward Emerson Barnard published a long paper (Barnard 1889) about his observations of Jupiter between 1879 and 1886 using a 5-inch refracting telescope, very similar in capability to the one at the Agricultural College, so that we know that much could have been seen by a skilled observer.

I would hate to think that there were no fun nights devoted to such observing with the Clark telescope, and I certainly envy those past observers their dark starlit skies that cannot be duplicated anywhere near the Lansing area today. Nevertheless, what it was actually like to peer through the telescope during an astronomy class of the early 1880s must remain in the realm of speculation unless we find some reminiscence from a long-ago stargazer who observed when Michigan itself had been a state for less than half a century.

*Photographs of Jupiter (1879), with its Great Red Spot, and Saturn (1885) from the 1893 edition of Agnes Clerke's History of Astronomy in the Nineteenth Century. The blue-sensitive photograph of Jupiter was taken by Andrew Ainslie Common and the one of Saturn by Paul and Prosper Henry.*

# III. Watching the Satellites: The Space-Age Arrives with Operation Moonwatch

Following the demise of the first campus observatory in the early 1900s, we jump ahead to a clear night in the late 1950s. Anticipation is growing atop the flat roof of what was then the Physics and Mathematics Building of the newly renamed Michigan State University. Close to a dozen observers wait for a moving point of light to cross the field of view of one of their telescopes. If and when it does, they will note its position in the sky and call out its appearance, so that the time can be accurately recorded. They are not watching for flying saucers or Soviet bombers[13]. They are volunteer members of the Lansing Moonwatch team, and they are helping to keep track of the orbits of the first generation of artificial satellites. It is an episode in campus astronomical history almost forgotten after the passage of six decades.

In 1955, the United States decided to launch a satellite as a part of the scientific endeavors associated with the International Geophysical Year (actually a year and a half, extending from July 1, 1957 until December 31, 1958). After deliberation, it was decided that the first

---

[13] In the early 1950s there was actually a corps of ground observers who did watch for Soviet bombers under the aegis of Operation Skywatch.

satellite would be Vanguard 1, a 3-pound, 6-inch sphere from which six small antennae protruded. However, at the time of the IGY, finding a satellite once it was launched, and determining the elements of its orbit, were not simple tasks. Radar of the time was not adequate for this purpose, nor were there yet photographic camera networks that could do the job. Radio could help, but only if a satellite were actively broadcasting a signal. Some way was needed for quickly finding approximate positions of satellites. Those positions would allow the determination of orbits adequate to tell the planned Baker-Nunn satellite-tracking cameras where to point and when to observe.

Fred Whipple, director of the Smithsonian Astrophysical Observatory (SAO), thought that amateurs – citizen scientists -- might be able to take up the slack. The Moonwatch Program (alternatively, Operation Moonwatch) was begun, with J. Allen Hynek, Armand Spitz, and Leon Campbell, Jr. initially handling much of the organization and communications with prospective Moonwatch teams. An excellent overview of the Moonwatch program can be found in the book by W. Patrick McCray (2008).

Plans for the new program, and a call for volunteers, were announced in 1956. Soon copies of the *Bulletin for Visual Observers of Satellites* began to be included in issues of *Sky and Telescope* magazine. By September of that year, some 35 Moonwatch stations, each with 10-15 members, had been organized. The first issue of the *Bulletin* stated the purpose of the program: "The primary objective of the visual program is to make sure that an observable satellite will not pass over a station without being observed with acceptable accuracy." Lansing, however, did not yet have a station of its own.

A Moonwatch station in mid-Michigan would start with two strikes against it. First was the cloudy weather, which not-infrequently hid the stars, especially in winter when lake effect clouds could blanket the skies for days at a time. Second was the northern latitude of Lansing, just under 43° north. The initial United States satellites were likely to be launched into orbits that did not carry them overhead at that latitude. Observers in Michigan might have to look to the south and hope that the distant satellites would be bright enough to see. The first Moonwatch *Bulletin* also warned that many enthusiasts of the fledgling space program would not have the skills or the time to contribute effectively to the team at a Moonwatch station. However, the Lansing effort was begun by amateur astronomers, which meant that it started with a corps of prospective observers who were already familiar with the night sky.

*Bulletin* number 1 explained that Moonwatch stations would not be lone-wolf operations. Diligent leaders would be needed to organize observing and report observations. Two Lansing amateurs, Harris

Dean (1918-1993) and Meinte Schuurmans (1899-1981), volunteered to lead the Lansing Moonwatch station, with the former offering to assume its directorship and the latter the deputy directorship.

Dean was in touch with the headquarters of Moonwatch operations in Cambridge, Massachusetts, before 1956 ended. The Lansing *State Journal* for January 1, 1957, reported that the local astronomy club, the Lansing Amateur Astronomers, in cooperation with the Michigan State University Physics Department, had begun organizing a Moonwatch station. Before that winter ended, a team of volunteer observers was being trained how to detect and record the passage of a satellite.

On March 7, 1957, Dr. J. Allen Hynek, Associate Director in charge of the satellite tracking program at the SAO, sent Dean news of the formal registration of the Lansing Moonwatch station. Hynek commented particularly upon the duties that Dean was undertaking: "I recognize that as team leader you will carry a great responsibility whose only reward is a deep appreciation of your efforts and the knowledge that your group is engaged in a scientifically significant venture." He also suggested that Dean should feel free to "contact your local newspapers, radio and TV stations to inform them of the contents of this letter. After all, you do have some news that is worth telling."

What was required to establish a Moonwatch station and how would one operate? The observers needed telescopes appropriate to their task. Moonwatch telescopes were not entirely standardized, but were often wide-angle, low-power instruments, similar in power to binoculars. To ease neck strain, the telescopes were often equipped with mirrors that allowed team members comfortably to see the sky while looking downward. Several companies were soon offering "satellite telescopes" for sale that would meet the criteria for Moonwatch observing.

In addition, a station would need a short-wave radio for receiving official government time signals. A tape recorder would permit simultaneous recording of radio time signals and alerting shouts from observers.

The figure below shows a model of a Moonwatch station that was constructed by Frank McConnell for the SAO. Observers with small telescopes are lined up, north-to-south, on each side of a central mast. Not all stations followed this arrangement, but the Lansing station would use it as a guide.[14]

Watchers on either side of the central mast were arranged along the north-south meridian. Their telescopes would point to the junction of the mast and a

---

[14] Rodney Dean and Marshall Dean, sons of Harris Dean who often accompanied their father to the Moonwatch station, recall that it was arranged much like McConnell's model. Phone conversations 7/17/19 and 7/18/19.

north-south crossbar. Each telescope would be located at a different, specified, distance from the mast, and thus would be watching a different altitude in the sky. The intention was that their fields of view would form an "optical fence", catching any satellite passing through that was bright enough to see.

*A model of a Moonwatch team in action. Courtesy of the Smithsonian Astrophysical Observatory Archives. Image number MAH-44383A.*

When a satellite was sighted by an observer, he or she would call out when it entered the field of view, crossed the center-line of the field, and exited. Times could later be noted from a tape recording, although some Moonwatch observers used stopwatches instead. The star field through which the satellite passed and the satellite's brightness would also be noted. These observations located the satellite in the sky at a particular time and, if enough observations were reported to Moonwatch headquarters in Cambridge, its orbit could be calculated. Sometimes even rough observations through a pair of binoculars might be useful in locating a satellite, but more accurate measurements were preferred. Procedures for making Moonwatch observations are described in the book by Neale E. Howard (1958).

Moonwatch teams ordinarily had to build or buy their own equipment. The price of a Moonwatch telescope does not seem high today. The Edmund Scientific Company (Edscorp) in Barrington, New Jersey, sold one for $49.50. It had an objective lens about 50-mm in diameter and, with a magnifying power of 5.5, provided a 12° field of view. However, $49.50 in 1957 is the equivalent of about $400 in 2019, a non-negligible sum and more than an equivalent instrument would probably cost today[15].

Lansing station leader Harris Dean stepped in to make involvement with Moonwatch more affordable. The Dean and Harris Ford dealership, with which Harris

---

[15] Original Edscorp satellite telescopes in great condition can, however, now sell for more than $400.

Dean was associated, provided funds to purchase a significant portion of the needed equipment. As a start, six of the Edscorp Moonwatch telescopes were purchased. Later, the team would supplement the Edscorp telescopes with the purchase of five Tasco 7-power, 40-mm instruments. In a letter dated May 3, 1957, Dean was able to report to Moonwatch headquarters that the Lansing station was now well equipped.

Besides equipment, a Moonwatch team needed a location where they could quickly set up for observing when a satellite passage was anticipated. Moreover, the station needed to be located where city lights would minimally brighten the night sky. The Lansing station's location, the Physics and Mathematics building, was built in 1949 and designed with a flat southern roof for astronomical observing, accessible from the second floor.

That rooftop was still used by students for occasional astronomical observing sessions as late as 2002, after which the Physics and Astronomy Department moved to the Biomedical and Physical Sciences Building. However, light pollution increasingly illuminated its skies in later decades. In the late 1950s, the sky was still dark enough for the site to be used by the Moonwatch team. This was Lansing station A (also known as station 34) in the parlance of the Moonwatch program. A few observations would also be made from

Harris Dean's home observatory, west of Lansing, which would be known as Lansing station B.

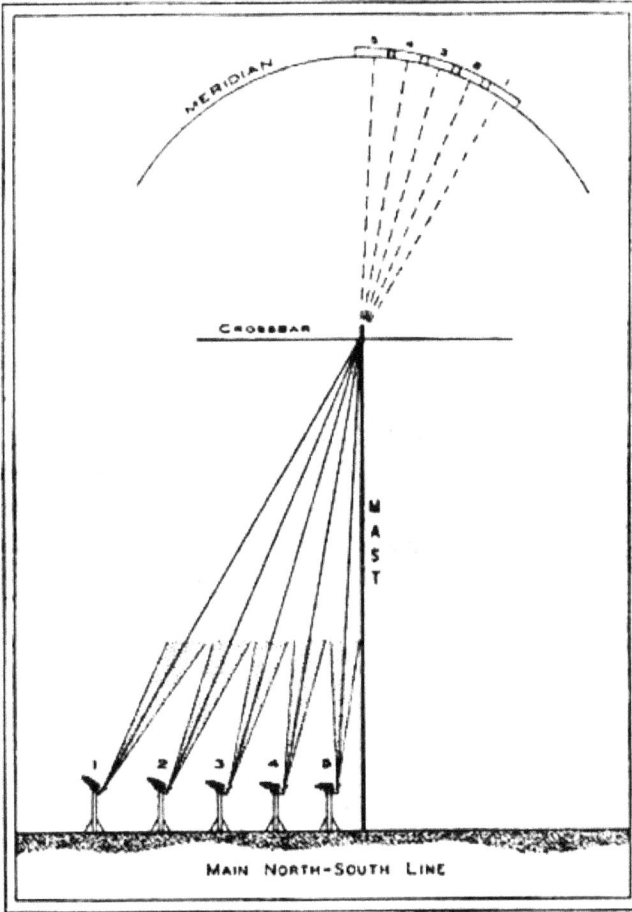

*This figure from the second issue of the Bulletin for Visual Observers of Satellites (October, 1956) shows how Moonwatch observers could set up an "optical fence" to catch crossing satellites. Courtesy Smithsonian Astrophysical Observatory Archives.*

*One of the Lansing Moonwatch station's Edscorp satellite telescopes. Note that the angle of the telescope to the horizontal could be adjusted. Photograph courtesy of Cameron Dean.*

The Lansing Moonwatch station as originally conceived had telescopes only on the north side of the mast, looking to the south, where it was expected the first United States satellites would be seen. However, as early as March 14, 1957, the possibility that Soviet satellites might someday need to be observed had apparently been broached in a phone conversation between Dean and Leon Campbell, Jr., the SAO's supervisor of station operations. Soviet satellites might be in orbits that would cross into the northern sky as viewed from mid-Michigan. It would in fact be the Soviet satellites that would end up being the main targets of the Lansing team, requiring observations to the north as well as the south and telescopes on both sides of the mast. Unfortunately, I have not been able to discover any photograph of the Lansing Moonwatch station set up at its rooftop observing location.

Correspondence in the SAO archives does not list the entire membership of the Lansing Moonwatch team. We do, however, on two occasions have listings, quite possibly incomplete, of some team members from the *State Journal* newspaper. Besides director Dean, and deputy director Schuurmans, the Lansing Moonwatch team initially included T. A. Louden, Wesley Clark, John Inman, Robert Rood, Richard Yarger, and Fred Hammond. No women were included in this first listing of team members. That was not extraordinary, as most Moonwatch teams were predominately male.[16]

---

[16] A letter from Harris Dean in the SAO archives gives the name of one more member, Bryant Pocock. Rodney and Marshall Dean, Harris's older sons, also served as observers, as noted below.

*The flat southern roof where the Moonwatch station was located is shown in this 1974 aerial photograph of what was, by then, the Physics and Astronomy Building. Structures on that roof postdate Moonwatch activities. MSU Historical Aerial Imagery.*

Because satellites in low orbits are most often seen within a couple hours of dusk or dawn, being a member of the Lansing Moonwatch team implied a willingness to rise in the wee hours of the morning or to rush off to observing after an early dinner. The only other Moonwatch station in Michigan appears to have been station 33, located near Detroit. The Lansing Moonwatch station appears to have been ready for action by mid-1957. However, for months there were no satellites to watch.

While the Moonwatch team practiced and waited, Harris Dean was asked to give talks to various civic groups about satellites and the Moonwatch program. These had to be improvised at first, since the Moonwatch program did not at the beginning have visual aids to lend to speakers. Giving such talks would become a regular occupation for Dean over the next few years, so that preparing and giving presentations must have posed a considerable burden for him. For a few years, Dean was the local go-to person for the Lansing *State Journal* regarding information about satellite sightings.

However, largely unbeknownst to the members of the Lansing Moonwatch group, plans were being laid in the Soviet Union to place a satellite into orbit before the launch of the first U.S. satellite. Sputnik, the first Soviet satellite, would be spherical, 23-inches (58-cm) in diameter, with a mass of 184 pounds (83.6 kg). That would make it much larger than the 6-inch and 3-pound Vanguard 1.

For the Moonwatch team, everything changed on October 4, 1957, when the Soviet Union announced that it had launched Sputnik. Most people in the United States were taken utterly by surprise, though perhaps they shouldn't have been. News that its cold war adversary had launched a satellite created great excitement and alarm in much of the country. Newspapers, such as the Lansing *State Journal*, would often use the loaded words "Red Moon" to refer to Sputnik and its successors.

Two objects were initially in orbit. There was the Sputnik I (alternatively Sputnik 1) satellite itself. It was rather faint and difficult to spot with the unaided eye. In addition, the larger core rocket booster that had helped put Sputnik I into space also orbited the Earth. It was much larger than Sputnik I and was bright enough to be readily seen with the naked eye when observing conditions were favorable (as when I saw it from Connecticut, a couple weeks after its launch).

Word of the launch of Sputnik I reached SAO Moonwatch headquarters late on a Friday afternoon, as staff were departing for the weekend. A harried staff quickly dispatched alerts and Moonwatch teams around the world were rushed into action. The professionally-operated Baker-Nunn camera stations that were intended to photograph satellites were not yet in operation. Much depended on the amateurs.

The Lansing Moonwatch team was keeping watch before dawn on October 5, but they did not spot anything that might have been the new satellite or its brighter rocket booster. Lansing area radio buffs listened for Sputnik's beep-beep signal. The first observations of Sputnik I by a Moonwatch team did not come until

October 8 from Australia and the first Moonwatch observations from the United States came on October 10 in Connecticut.

The orbit of Sputnik I had an inclination of 65°, large enough to carry the satellite overhead for the Lansing area and requiring modification of the Lansing station's intention to focus on the southern sky. However, Sputnik I was not favorably placed for observations from Michigan at either dusk or dawn in the weeks immediately after its launch. The Sputnik passages observed by the New Haven, Connecticut, Moonwatch team on October 10 and 11 were located too far to the east for viewing by the Lansing group. Between its orbital path and clouds, it appears that the Lansing Mooonwatch team may not have caught the first satellite. At least, I have been unable to discover any reports that it saw Sputnik I, and such an event would have been likely to have been commemorated in the *State Journal*.[17]

The luck of the Lansing team changed when, within a month of the launch of Sputnik I, the Soviets repeated their success. Sputnik II, carrying the sacrificial dog, Laika, was sent into space on November 3. The scientific payload went into orbit attached to its

---

[17] Rodney and Marshall Dean both recalled searches for Sputnik I, so it is possible that the team witnessed it at some point, even though I have not found documentary evidence for such a sighting.

booster core, making Sputnik II a bright target for naked eye observations. It was seen from East Lansing before sunrise on November 6 with an apparent magnitude of -2, brighter than the brightest stars, an observation which Dean quickly phoned to the SAO. The *Detroit Free Press* for Thursday, November 7, was able to report "Sputnik II was sighted all across the country Wednesday, giving scientists information for a fix of its orbit around the earth." However, the luck of the Lansing team had not entirely changed, as attempts to obtain observations that winter were often frustrated by poor weather.

Though Harris Dean's two oldest sons, Rodney and Marshall, were children at the time, they were already skilled enough in the ways of astronomy to man two of the Moonwatch telescopes. Six decades later, Rodney remembered the thrill of having a satellite appear in the field of view of his telescope on one occasion, while Marshall recalled his disappointment that the satellite favored his brother rather than himself! This reminds us that, even when a Moonwatch team successfully saw a satellite, most of the team members would not see it cross through the field of view of their assigned telescope.[18]

---

[18] Phone conversations with Rodney and Marshall Dean on 7/17/19 and 7/18/19, respectively.

Following the explosive failure of the attempt to launch the first Vanguard satellite, the United States succeeded in putting a different satellite, Explorer 1, into orbit on January 31, 1958. Unfortunately for the Lansing team, Explorer 1 was relatively small and faint, and its orbit did not reach sufficiently far north to make it a good target for the Lansing station. The same was true for other American satellites launched that year.

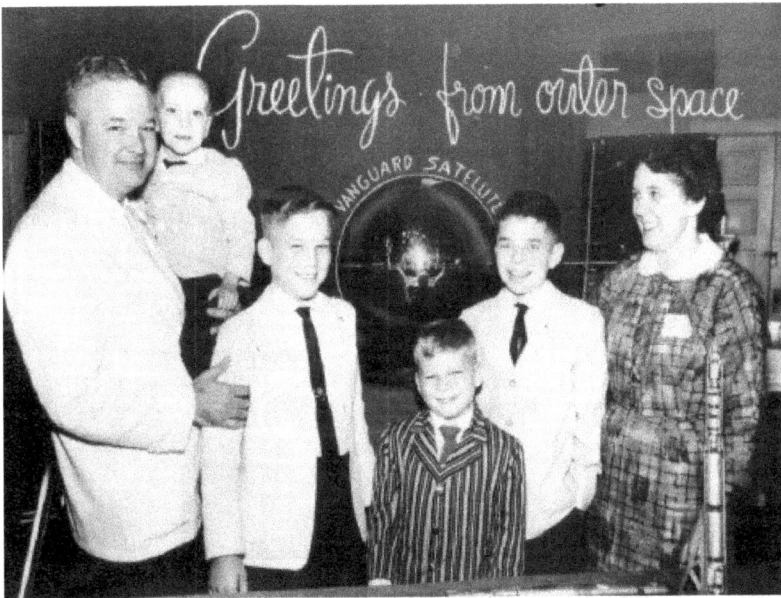

*Harris D. Dean and family at the Grand Hotel, Mackinac Island, in August, 1958, where he spoke about the space program. Courtesy of the Dean family.*

The team had better success when the only Soviet satellite launched in 1958, Sputnik III, went into orbit on May 15. Lansing observers caught Sputnik III passing over as early as the evening of May 16. In those days, each new satellite garnered significant attention by the news media and, the next day, Dean reported to the *State Journal* that "the satellite appeared in the sky directly west of his position on top of the physics building of Michigan State University."

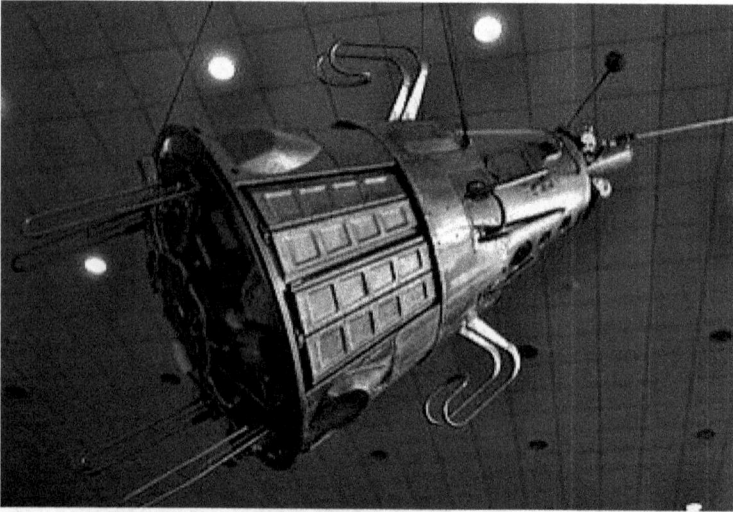

*Sputnik III, shown here, and its booster were targets of the Lansing Moonwatch team. Courtesy of Wikimedia Commons and Енин Арсении. Creative Commons Attribution-Share Alike 3.0 License.*

As with Sputnik I, more than a single object was initially in orbit. Besides the conical science package, the tumbling rocket booster was also placed into orbit. As it tumbled with a period of eight or nine seconds, it sometimes appeared as bright as the brightest stars and sometimes only as bright as a star of the fourth magnitude. Sputnik III and its rocket were the objects best studied by the Lansing Moonwatch team. Lansing observations were usually of what was then called 1958 delta 1, the brighter carrier rocket.

Observations by Moonwatch teams provided information not only about the orbital elements of satellites, but, indirectly, also about the density of the tenuous atmosphere high above the ground and the precise shape of Earth's equatorial bulge. Moonwatch teams also participated in so-called "deathwatches", watching for the fiery reentry of satellites as they plunged back into the denser atmosphere. Reentry times could not then be predicted with any accuracy, and even today can be associated with substantial uncertainties. One of these deathwatches involved the reentry of Sputnik II. Its eventual fall on April 14, 1958, came too far east for the Lansing team, but its demise was caught by three east coast Moonwatch groups.

The extended International Geophysical Year drew to a close at the end of 1958. By then the main contributions of the Lansing Moonwatch team had been made. On January 28, 1959, the *State Journal* reported that "at a meeting of the Lansing Amateur Astronomers February 7, in the State Journal lounge, awards will be given from the Smithsonian Astrophysical Observatory, Cambridge, Mass., to moonwatchers and a local backer. A certificate of appreciation will be presented to the Dean and Harris Company, local automobile dealers, for providing equipment for the satellite tracking program, in addition to citations to the following members of the Moonwatch group: Wesley Clark, Dr. John Inman, Robert Elliott, Mrs. Neil Newman, Robert Rood, Meinte Schuurmans, Fred E. Hammond, Mrs. Bette Smith, Richard Yarger, and Harris Dean." By then the Lansing team appears to have gained at least two women members.

A letter from Leon Campbell, Jr., dated March 26, 1959, congratulated the Lansing Moonwatch team on reaching "standard" status, given to teams which had contributed significant observations.

The number of Moonwatch teams climbed to a high of 230 in 1958. Of those, 128 were in the United States and some 80% of those made useful observations. With the ending of the IGY, participation in the Moonwatch program gradually diminished. Satellites became more familiar, and other tracking methods began to supplant the Moonwatch teams. The Baker-Nunn tracking network had become fully operational in July, 1958, and there was less need for an abundance of Moonwatch stations. Those that remained in operation began to be tasked with observing fainter satellites, requiring larger telescopes and darker observing locations.

Health issues forced Harris Dean to withdraw as leader of the Lansing team in the fall of 1959. The Lansing Moonwatch station was placed on reserve status as of April 1, 1960, and was retired from the program in June of that year. Although the Lansing Moonwatch station was in operation for only a few years, its members made a meaningful contribution in the exciting pioneering days of the Space Age.

The MSU Campus Observatories

52

# IV. Birth of the Second Campus Observatory

Half a century after the demise of the 1880 campus observatory, the University of Michigan remained the only institution in the state having an astronomical observatory with a research focus. The launch of Sputnik in 1957, which inaugurated the space race and activated the Lansing Moonwatch team, also fostered an increased general interest in astronomy. Abrams Planetarium opened on the campus of Michigan State University in 1964. However, by the mid-1960s, Michigan State University still had not constructed a replacement for its long vanished original campus observatory.

In the early 1960s, astronomy was taught at MSU within a Department of Physics and Astronomy, but at that time no astronomer was actually on the department's faculty. In 1966, astronomy was separated from physics and a new Department of Astronomy was established, chaired by Albert P. Linnell (1922 – 2017), who came to MSU from Amherst College to direct the new program.

By this time, research telescopes were no longer instruments to look through; photographic emulsions and photoelectric photometers had replaced the eye as light detectors. Linnell's 1950 Harvard PhD thesis dealt with photoelectric photometry and his first research paper reported photometry of UX Ursae Majoris, an eclipsing variable star with nova-like properties. He would maintain an interest in binary stars throughout his career, with his final paper on that subject being published 63 years later in 2013. When he came to MSU, there was much to do establishing the new department, but a new campus observatory and photoelectric photometry were very much on his mind.

In 1965, even before the Astronomy Department was formed, Thomas H. Osgood, then director of Abrams Planetarium and a prominent faculty member in the Department of Physics and Astronomy, had written to President John Hannah, advocating that a student observatory be built and operated in conjunction with the planetarium (Suelter, 2008, p. 227). Nothing came directly from this proposal, but Osgood would play an important role in encouraging the creation of the independent Astronomy Department.

Linnell, when he came to MSU, had a research observatory in mind, rather than one focusing solely on student education. In an October 5, 1966, letter to Provost Howard Neville, Linnell advocated the purchase of two telescopes. One, a 16-inch reflector, would be housed in an observatory on campus, within easy reach of students. The second, a 40-inch reflector, would be placed in a darker location farther from campus lights. Linnell proposed buying the telescopes from Boller and

Chivens corporation, builders of many research-grade telescopes in the 1960s (Suelter 2008, pp. 227-228). In the end, only a single observatory would be built.

Perhaps it is worth considering why it was decided to put a new research observatory in often cloudy Michigan. Of course, in the years before the internet, having an observatory nearby greatly facilitated student and faculty involvement with observations and instrumentation, especially for undergraduates. A local observatory also eased support expenses and kept travel expenses low. An observatory on land the university already owned would further cut costs, and, moreover, a local observatory would keep the work of the astronomers before the eyes of the administration, something potentially of value to a young department.

It is also worth remembering that, at that time, Michigan State University would by no means be alone in operating an astronomical observatory in the Midwest. A number of well-established research observatories were active in Michigan and nearby states throughout the 1960s. Observations were still being conducted at the historic Yerkes Observatory in Williams Bay, Wisconsin. The forty-year old Perkins Observatory in Delaware, Ohio, had lost its 69-inch telescope to Arizona, but replaced it with a 32-inch reflector. The Pine Bluff Observatory was opened by the University of Wisconsin in 1958 at a location 24-km west of Madison. Case Western Reserve University operated the Burrell

Schmidt telescope from sites near Cleveland from 1941 until its removal to Kitt Peak in 1979. The Lindheimer Astronomical Research Center, built in 1966 on the shore of Lake Michigan at Northwestern University's Evanston, Illinois, campus, operated 1-m and 0.4-m telescopes until the mid-1990s. The Ritter Observatory of the University of Toledo, with a 1-meter telescope, was established in 1968 amidst the lights of the Ohio city for which the university was named. The University of Illinois opened its Prairie Observatory, also with a 1-m telescope, farther from city lights in 1969, and kept it at its original Illinois location until 1981.

By the mid-twentieth century, the University of Michigan had abandoned the city of Ann Arbor as a site for its research telescopes, but it was still siting telescopes within the state of Michigan. The University of Michigan inaugurated the Curtis Schmidt telescope at a relatively dark Michigan site in 1950. It would be moved to Cerro Tololo Interamerican Observatory in 1967, but, even then, the University of Michigan was not finished with local observing. It acquired a 1.3-m reflector in 1969 that would remain in Michigan until its removal to Arizona in 1975. Michigan State University was thus far from unique when it built an observatory in the Midwest.

Plans for the observatory advanced in the next couple years, and the rationale behind the choices made can be seen in a proposal to the National Science

Foundation asking for partial support for an observatory, dated January 10, 1969. Linnell and a second Astronomy faculty member, Thomas R. Stoeckley, who had been hired in September, 1967, were the principal investigators on the proposal. By the time that the proposal was submitted, the observatory site and building layout had already been decided, and a 24-inch Boller and Chivens telescope had been ordered.

It was decided to keep the observatory within 10 miles of the central campus so that it was accessible to students. Inquiries to several existing observatories in Michigan, Ohio, Wisconsin, and Illinois indicated that, even under the worst likely observing conditions, 500 hours of quality observing time might be expected in the course of a year for a single telescope, and that more observing time might often occur. That turned out to not be far off the mark. For example, between 1 October, 1976, and 30 September, 1977, the 24-inch would be used for 714 hours on 180 nights (Stoeckley 1978).

It is worth quoting from the 1969 proposal to show the overall thinking behind the observatory design:

*"The observatory building as it appears in this proposal is the end product of more than a year's consideration, improvement, and modification of design.*

*"It had been determined near the beginning of the observatory project that two telescopes should be included. One would be a relatively small instrument usable for a variety of purposes including direct photography, photometry*

*with a Cassegrain photometer, and spectroscopy with a coudé spectrograph. The larger instrument would be useful for higher dispersion spectroscopy and occasionally for other purposes."*

The proposal went on to outline five planned areas of research for the new observatory:

A) *Multichannel simultaneous photometry of eclipsing binaries;*
B) *Spectral classification of eclipsing binaries;*
C) *Radial velocity curves of eclipsing binaries;*
D) *Stellar rotation velocities of single stars;*
E) *Stellar radial velocities of visual and eclipsing binaries.*

Provision was made for other potential uses of the telescope – for example, laser ranging of the moon. Single-channel photoelectric photometry of eclipsing variables, which would be the observatory's main early scientific program, is rather neglected in the proposal, presumably because the proposal emphasizes those projects for which NSF support was especially needed.

To accomplish these research goals, it was proposed that the new observatory be home not only to a 24-inch (61-cm) reflecting telescope but also to a larger innovative coudé feed mirror telescope. A 66-inch (1.55-m) flat mirror, located on a pier outside the observatory building, would send light to a 40-inch (1-m) parabolic mirror, also on an external pier, that would in turn feed light to a spectrograph inside the observatory. The proposal noted that observing conditions were best in East Lansing in the late summer and early fall, times when the seasonal monsoon often made observing difficult at Kitt Peak National Observatory in Arizona.

The NSF proposal was not funded, and the external coudé feed telescope was never actually implemented. Nevertheless, the observatory was built during 1969 on an agricultural area of the south campus, north of the intersection of Forest Road and College Road (latitude 42 degree 42 minutes 23 seconds; longitude 84 degrees, 29 minutes, 00 seconds west, elevation 866 feet above sea level – values from Google Earth, 2017).

It was a two-story building, topped by a dome purchased from Astro-Tec in Ohio. The 24-inch Boller and Chivens telescope, purchased with university funding, remains the main instrument of the observatory. It was installed on an equatorial mounting in the observatory in February, 1970. The total cost of the 24-inch telescope was about $125,000 dollars, while the observatory as a whole cost about $660,000 (MSU press release, 1970).

*A winter view of the MSU campus observatory not long after the observatory was built, looking to the southeast. Note the shielded parking lot lights, which could be turned off during observing. Observatory collection.*

The firm of Kenneth C. Black Associates served as architects for the new observatory, although they drew their plans with a great deal of input from the astronomers and other university personnel. Charles Featherly Construction Company did the general construction work. The new observatory had room for more than just a telescope. It contained office and storage rooms, space for an electronics shop, a room with a kitchen that could be converted to a temporary bedroom, a darkroom for photographic work, space for future development of the coudé feed telescope, and a room to house a computer for controlling the telescope and instrumentation. Such extensive use of a computer was still relatively new at an observatory and will be discussed further in Chapter V.

The telescope was located at the top of the observatory building, but on a pier separate from the building structure to prevent vibrations from the building from shaking the image in the telescope. The pier itself was supported by concrete caissons sunk deep into the ground.

Keeping images from the telescope as sharp as possible was a matter of concern from the beginning. Placing the telescope in a dome high above the ground might help seeing, but placing it amidst a larger, heated building could hurt image quality. To alleviate this problem, a complex heating and cooling system was designed, with the idea that the system would always maintain the dome and telescope at the outside ambient temperature. This air conditioning system would also maintain stable temperatures in a planned second floor spectrograph room and in the computer room. As we shall see, the air conditioning system would soon cause trouble.

No elevator was installed, but a crane attached to the top of the dome enabled heavy equipment to be lifted through trap doors that could be opened in the observatory flooring.

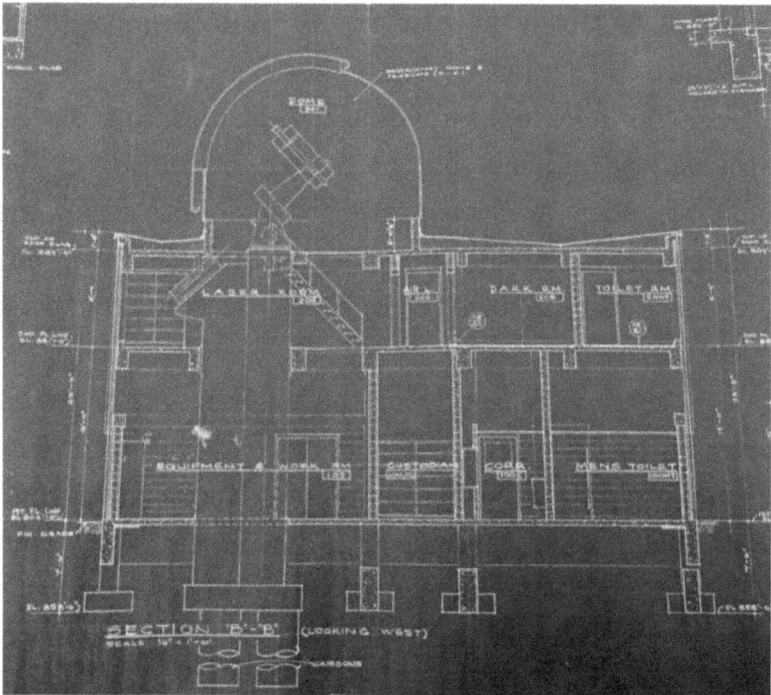

*A portion of the original blueprints for the observatory building, showing a south-to-north slice through the dome and building. Observatory collection.*

The Boller and Chivens corporation constructed more than twenty 24-inch telescopes during the 1960s and 1970s (see http://bollerandchivens.com/). Few, however, were constructed so as to offer the option of a coudé focus, as was the case with the MSU telescope. Moreover, the MSU telescope was designed from the start to be positioned with encoders, enabling eventual computer control. Thus, the MSU telescope did not have the analog position dials often seen on the skirts of

Boller and Chivens telescopes made at that time. The original encoders had to be reset if the power to them was turned off or lost, requiring that a bright star be manually centered and its coordinates entered.

The telescope was of the Ritchey-Chretien design with an f/3.0 primary mirror made of low-expansion Cer-Vit material. Two alternative top ends could be installed on the 24-inch. One had a secondary mirror giving a final f/13.5 focus. The second had a secondary mirror with two aluminized surfaces that could be flipped to change the focal length. The first side provided an f/8 focus, while the second provided an f/34.6 beam for the coudé focus. When used at the coudé focus, a third mirror would send the light down the polar axis to the floor below. A 6-inch (15 cm) f/15 refractor was attached to the 24-inch for use as a guide telescope, and there was a 50-mm finder telescope as well. A console permitted operation of the telescope from the dome.

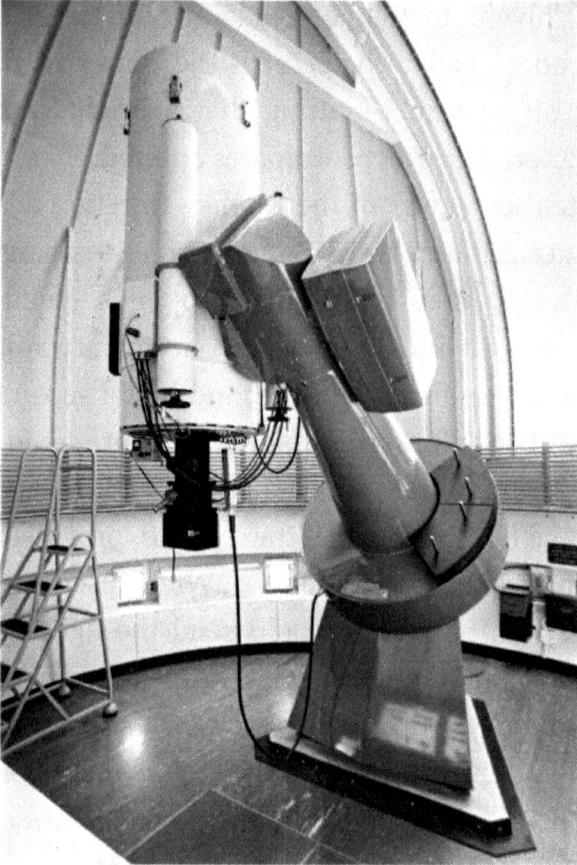

*The 24-inch telescope with photoelectric photometer. This photograph, courtesy of Robert Miller, dates to the early 1970s.*

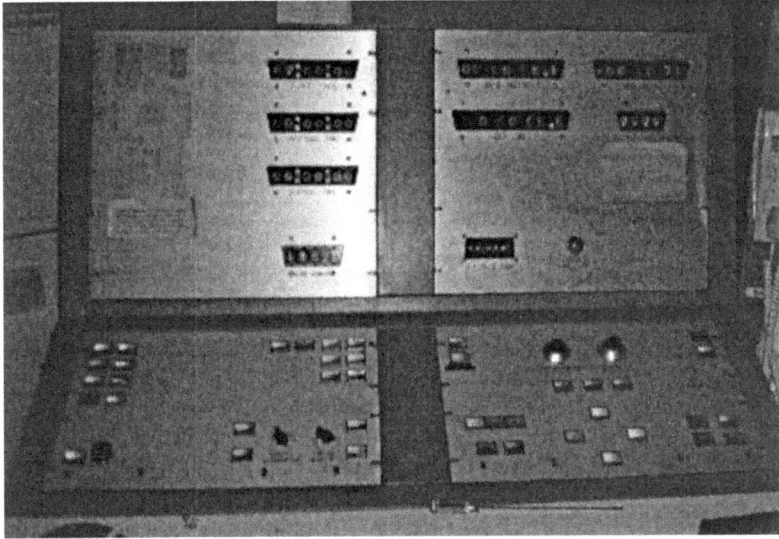

*The large console on the dome floor. Positions were read from nixie tube displays, the changing red numbers of which were always a source of amusement when small children visited the observatory. Author's photograph.*

The MSU Campus Observatories

## V. The 1970s: Progress but Problems

Though we have no quantitative estimates of sky brightness at the observatory during its early years, those who used the telescope in the 1970s agree that the sky was much darker than it would later become. The observatory's surroundings were agricultural, and the barns and sheds in the neighborhood created little light pollution (though they did occasionally result in some olfactory pollution). Moreover, the area to the east and southeast was far less developed than it would become during the 1980s, when the southern portion of Okemos Road, and the Okemos-Jolly Road intersection would begin to see substantial development. Even to the north, toward the main campus, there was less light than would later be seen, although there was already some light pollution in that direction and also to the west, toward the city of Lansing. Smoke from the coal-burning campus power plant could occasionally drift above the observatory, but prevailing winds generally carried it away.

*An aerial view of the observatory in 1980, when its neighbors were still barns and other agricultural buildings. (MSU, GIS Historical Imagery). Remaining barns to the west would be removed and Forest Akers Golf Course extended soon after the observatory's 1986 reopening. North at top. East at right.*

A press release in July, 1970, announced that the new observatory was finished and ready for use. In fact, there was much for the MSU astronomers still to do before the main scientific programs of the observatory could begin. The telescope was only ready to be used visually and for direct photography. With a glass corrector lens slid into place before the plate holder, 4 x 5-inch plates gave good images over more than a 2-degree field of view. However, no spectrograph was in place, nor was there yet a system for carrying out photoelectric photometry.

During the initial year of operation much effort was put into taking photographs to test the optics and collimation of the telescope. It was not long before photographs of comets, nebulae, and star clusters were being obtained.

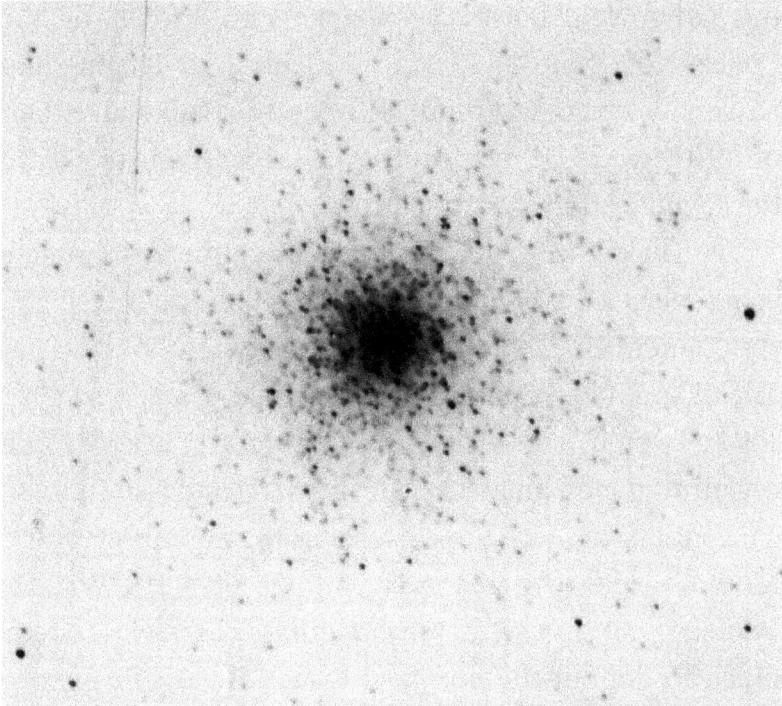

*This figure shows the central portion of a photographic plate of the globular cluster M13, taken on June 26, 1970. The 20-minute exposure on a Kodak IIa-O emulsion was unguided. The star images show a slight elongation, but the telescope tracking appears to have been working well. The plate was later used in studies of the variable stars in M13. Observatory plate archive. Scan courtesy of Wayne Osborn.*

The official dedication of the observatory came a year after first light, on May 8, 1971, coinciding with meetings at MSU of the Midwest Astronomers group and the Ohio Neighborhood Astronomers. Dean of the College of Natural Science Richard Byerrum and university President Clifton R. Wharton spoke on behalf of MSU, while Caltech professor Jesse Greenstein gave the keynote address (Linnell 1972).

The dedication did not mean that the astronomers could relax. An even larger meeting, the 138th meeting of the American Astronomical Society, was held at Michigan State University in August, 1972. August, 1974, would see the Astronomical League, an organization of amateur astronomers, hold its annual convention at MSU, with visits to the observatory. Moreover, in 1973 the AST 437 Observatory Practice Course began to use the observatory. There was indeed much to do, but the hiring of Ernest Brandt in 1972 to work as a technician at the observatory was a help.

A major effort was put into designing and building an automated system for photoelectric photometry of variable stars. Such an automated system was rather novel in 1970 when often all of the steps in making an observation were accomplished manually. To automate the photometry, a computer was needed.

A computer room had been built into the observatory in anticipation, but, without an NSF grant, funding the computer proved difficult. Professor Linnell wrote that the "computer was obtained from contingency funds in the observatory construction budget", but those funds were insufficient to immediately purchase the full configuration needed. A Raytheon 706 computer was purchased, but additional components had to be added over several years as funding permitted. These would eventually include tape drives, a cathode ray tube display in the dome, a paper tape reader, and a punch card reader. An interface allowing the computer to control the telescope and instruments also had to be designed and built.

The photoelectric photometer itself, based on an f/13.5 photometer at Kitt Peak National Observatory and cooled with dry ice, was built at MSU and completed late in 1972.[19] Work on the automated photometry system began to occupy much of the time of Professor Linnell and new faculty hire Stephen J. Hill.

---

[19] Robert Miller reports that a device that made frozen $CO_2$ for the photometer exploded due to a pressure blockage. No one was hurt, but it could easily have been otherwise. Astronomy can be dangerous!

*This photograph from the early 1970s shows an erfle eyepiece on the 24-inch telescope. A plate/film holder could slide into place for photography. The tubes with cranks on the end hold weights that can be moved to balance the telescope. Courtesy Robert Miller.*

The July, 1970, report in the *State News* announcing that the new observatory was ready for use quoted Professor Linnell: "[The telescope] will not be used for general observation. There are so many important problems to be worked on that we can't tie up an expensive instrument like this just for visual observing." However, as is often the case, subsequent developments did not entirely conform to original

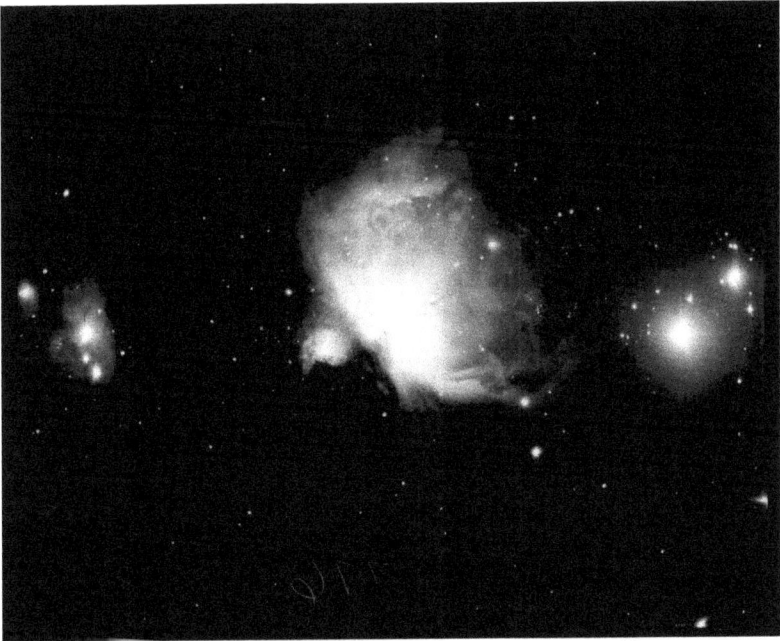

*This 30-minute exposure of the Orion Nebula was taken in 1977 with the 24-inch telescope using a photographic plate having a 103aF Kodak emulsion. North at left. Observatory plate archive.*

expectations. By June 1971, soon after its official dedication, the observatory was opened for monthly public viewing nights, thanks in part to the efforts of graduate student Robert D. Miller and faculty member R. Erik Zimmermann. On those occasions, the single narrow stairway from the second floor to the dome made (and still makes) an awkward passage for the public, but a dome with separate entrance and exit stairways would have required more complicated and expensive construction, and also a greater awareness that public viewing would be an important observatory use.

A few years after the observatory opened, Comet Kohoutek received extensive publicity, and many expected it to shine brilliantly as 1973 ended and 1974 began. Public interest was great and public observing at dusk was scheduled for January 12 and 19, although the comet's low altitude made observing with small telescopes more practical than with the 24-inch telescope. Nonetheless, on the 12th, a crowd of 500 people turned out at the observatory. The lines to see the comet moved briskly, but the comet set before all could see it. Alas, Comet Kohoutek proved to be far less bright than anticipated, disappointing many in the general public and giving astronomers worldwide something of a black eye.

Facilities available at the observatory increased when Herbert J. Rood joined the faculty in 1972. When Professor Rood arrived from Wesleyan University, he brought with him a Joyce-Loebl microdensitometer for scanning plates. That somewhat finicky instrument was installed at the observatory, where it was eventually used in several projects.

*Dedication of the new observatory in May, 1971. Caltech Professor Jesse Greenstein speaks, while Professor Linnell sits behind him with MSU president Clifton Wharton to the left of the podium. Observatory collection.*

*Photograph of Comet Kohoutek on an evening early in 1974, looking to the southwest over agricultural buildings near the observatory. Observatory collection, photographer unknown.*

*The observatory's Raytheon computer room in the mid-1970s. Observatory collection.*

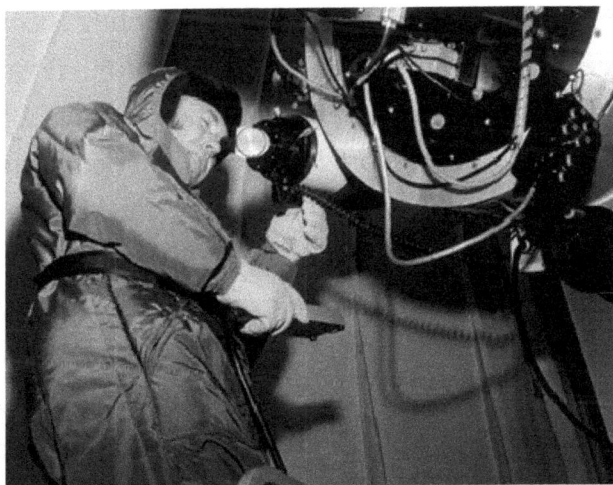

*Professor Linnell peers into the eyepiece of the 6-inch guide telescope on a cold night sometime in the early 1970s. Observatory Collection.*

Meanwhile Linnell, Brandt, and Hill struggled with completing the telescope-computer interface and developing a working automated system for high-speed photoelectric photometry. At the time, most photoelectric photometry systems required the telescope to be pointed, and the filter slide moved, by hand. Moreover, photometric readings were often output to rolls of paper, or even to a strip chart recorder, with the result that data reduction was time consuming and not easily handled directly by a computer. Given limited resources, making a computerized system that overcame those constraints was not a simple task. Professor Hill and graduate student Robert Miller were, however, soon taking photometric data to investigate non-variable stars within the instability strip.

The continued failure of the cooling system to work as planned was a cause of exasperation. It was also a cause for embarrassment when the Sunday, April 11, 1976, issue of the *Detroit Free Press* carried a story by David Johnston with the headline "$1 Million MSU Observatory Finds No Stars, Just Kinks." The story emphasized the failure of the expensive cooling system, but also criticized a lack of research publications from the observatory, the absence of NSF support for the research program, and complaints about student access to the observatory. It noted reports that "under certain conditions, it literally rains inside the dome". That was an exaggeration, but condensation could indeed be a problem when the dome was cooled.

Overall, it was not a pleasant article for Professor Linnell and the other astronomy faculty to read, nor one that would place the observatory in a good light before the public or the university administration. It must have been small consolation that the story included a nice photograph of the observatory from the west with a barn and cows in front.

Though the hoped for external coudé feed mirror for the new observatory was never built, the 24-inch telescope had its own coudé focus and Professor Stoeckley had not given up on implementing some form of spectroscopy at the observatory. In the late 1970s, using inexpensive parts, and doing much of the work himself with the assistance of Robert Miller, Stoeckley installed two spectrographs in what was originally planned to be the spectrograph room on the second floor. Light from the 24-inch telescope could be directed to a lower dispersion 70 angstroms per millimeter spectrograph or to a higher dispersion 13 angstroms per millimeter spectrograph. Given the low quantum efficiency of photographic plates, and the limited light gathering power of the coudé arrangement, exposures were long, especially for the higher resolution spectrograph, which was limited mainly to bright stars and planets. The spectrographs nonetheless proved to be effective teaching tools (Stoeckley 1977, 1978, 1979, 1980, 1981).

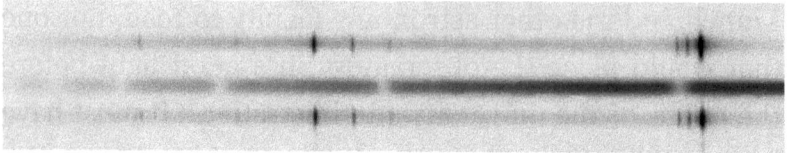

*A portion of the (negative) spectrum of Regulus showing the Balmer absorption lines of hydrogen as well as a comparison emission line spectrum. Obtained with the high-resolution spectrograph in 1978. Observatory plate archive.*

We now come to an unfortunate time in the history of the observatory. As the 70s progressed, disagreements apparently developed among some members of the astronomy faculty which led to ill-feeling. I lack adequate information to discuss all aspects of these difficulties, nor is there a need to do so, as some involved issues not connected with the observatory. However, the automated photometry system and the proper use of the observatory became subjects of differing opinions within the department.

Linnell, Hill, and Brandt (1975) had published a paper on the first automated photoelectric photometry system that was installed at the observatory. Professor Linnell found it necessary to make changes to the observatory electronics, changes that meant that this first photometry system no longer worked. He eventually installed a second automated photometry system capable of rapid cadence observations (Linnell 1981). Unfortunately, these changes were apparently one cause of the friction within the department. In 1979 a departmental review committee composed of astronomers from other universities and chaired by Alar Toomre visited MSU and arrived at conclusions critical of the revised automated photometry system, criticisms with which Professor Linnell did not agree. There is no need to hash out these differences here, nor is it possible now to establish who was right or wrong on every point. However, the upshot appears to be that Professor Linnell's further plans for the observatory were not carried out.

As the 1970s drew to a close, the photoelectric photometer could be operated using Professor Linnell's automated system, or it could be operated in a more manual mode if preferred (Stoeckley 1980). By this time financial troubles loomed at MSU. What role disputes within the department may have had on subsequent administration decisions regarding the future of the observatory is unknown (see Chapter VI).

These controversies should not obscure the fact that as the 1970s ended progress was finally being made on publishing scientific results from the observatory. Professor Linnell published photoelectric photometry of the W Ursae Majoris eclipsing binary star VW Cephei and studied its period changes, while faculty and students presented talks at meetings of the American Astronomical Society that were based upon data from the observatory. Walter Bisard and Wayne Osborn, professors at Central Michigan University, used photographic plates obtained with the 24-inch telescope to study variable stars in the globular cluster M13. Papers resulting from these efforts are listed in Appendix II. We shall, however, see in the following chapter that this progress would be brought to an abrupt halt as the next decade began.

––––––––––––––––

*Because few on campus today recall the early years of the second campus observatory, I have asked one person who does, Robert D. Miller, to recount his involvement with the observatory in the 1970s. His account follows.*

### Recollections of the Early Years at the MSU Observatory

### Robert D. Miller

My initial visit to the MSU campus and tour of the observatory began as I was completing requirements for a Bachelor's Degree at Indiana University. This visit occurred near the end of February, 1970, the day following the installation of the Boller and Chivens 24-

inch telescope. Boller and Chivens was a well-regarded builder of precision telescopes and accessories and was a part of Perkin Elmer Corporation.

I had become acquainted with Michigan State University a couple years prior to my visit through my involvement with the Fort Wayne Astronomical Society. One of the active senior members of the Fort Wayne club was Robert Stoeckley, who was teaching with the planetarium of a local college. Robert's younger son, Thomas, was the second member to join the new MSU Astronomy faculty. He later became its department chair. The Stoeckleys encouraged me to enroll in the graduate program in Planetarium Science at MSU.

The Fort Wayne Society has an excellent 12-inch reflecting telescope, that I, along with several other members, used extensively. That telescope provided valuable observing experience using what was, for that era, a relatively large and sophisticated amateur telescope. During that time, several club members and I cultivated an interest in building telescopes. Grinding, polishing and figuring telescope mirrors of high quality became something for which I developed a bit of a reputation. That skill helped some years later at MSU, when I produced the mirrors for the observatory's coudé spectrograph.

My study toward a Masters degree began in the summer of 1970. With teaching credentials already established, I proceeded by enrolling primarily in upper-level astronomy courses, which were my main interest. Not only was Tom Stoeckley an unusually gifted and respected faculty member, his helpful and attentive attitude toward students was legendary. Throughout this study, I was given a valuable graduate teaching assistantship through Abrams Planetarium. (Teaching morning classes after all-night observing, especially in winter, wasn't easy!)

At MSU, my interest in teaching and research at the new observatory developed quickly. During the telescope's first year, Tom and I used the new telescope extensively to test the optics and the telescope's star tracking. The telescope had an advanced optical system (a Ritchey-Chretien design) which included a field-flattening lens to enhance wide-field photography. We eventually demonstrated that this telescope was very useful for positional astronomy, with its well-corrected wide field on 4 x 5-inch photographic plates. The observatory has a Mann high-precision measuring machine, which we used to analyze the images.

In addition to evaluating the telescope's performance, during that first year or so many photographs were taken of galaxies and star clusters and a few comets. I made many of these photographs, but a number of students contributed. We made an effort to obtain photos of those galaxies in which supernovae were discovered. (I often remained on campus through term breaks in order to have unlimited access to the telescope.) One student, Claus Buchholz, even made a color photo of the Orion Nebula, using a color-separation technique which was an impressive accomplishment at the time, long before the advent of digital photography.

Photographic plates, while they can record extraordinary detail, have poor quantum efficiency. The best emulsions, designed for astronomy, captured less than two percent of the incoming light. At MSU, as at many observatories, experiments were conducted to improve that efficiency, such as chemical or physical treatments of the films or plates. Even obtaining a one percent improvement would be similar to getting a larger telescope, but much cheaper. As the cost of glass photographic plates was quickly becoming prohibitively expensive in the 1970s, more often than not, we resorted to using 4 x 5-inch film as a substitute. By comparison, nowadays, photographic films and plates have been replaced with electronic detectors, whose efficiency runs in the range of 50 to 70 percent. While electronic detectors are typically much smaller in size than traditional films or plates used in photography (therefore yielding a smaller field of view), their high quantum efficiencies and linearity far outweigh their smaller size. An interesting note: some of those early star cluster photographs are currently being used for variable star research.

In late 1970, faculty member Erik Zimmermann and I began a series of public observing evenings at the observatory. These eventually became a long-running program in which two evenings each month were set aside for public viewing through the 24-inch telescope. As we acquired a set of smaller, portable telescopes, we used them for observing from the area adjacent to the observatory, to better accommodate the large public interest in viewing the night sky. These public programs continue, as popular as ever, nearly fifty years later.

After receiving a Masters degree in 1972, I accepted a teaching job, becoming a planetarium curator in Florida. It took only a short time for me to realize this was not to be my long-range career choice. At the time, I was being interviewed for a position at a major observatory to continue research and development of photographic materials and techniques, but it appeared that funding for that position might become unlikely. It was time to consider Plan B! I chose to return to MSU in the fall of 1973, to further extend my preparation in astronomy -- in the event that the photographic research program would be funded -- and to find another career otherwise.

In the early 1970s, it was realized that the observatory needed a Cassegrain or coudé spectrograph. In order to design such a system, Tom Stoeckley developed a sophisticated optical ray-tracing computer program in which optical parameters, such as glass types and their properties of dispersions, sizes, and radii of curvature, were used to compute spot diagrams (the image a point-like optical source produced with the optical system); such diagrams quickly reveal design limitations.

A coudé spectrograph was actually built some years later, mostly by Tom Stoeckley and me. I made the mirrors for it, including a four-inch collimator and two 8-inch imaging mirrors, at f/5 and f/20. This was intended both as a training instrument and as a test bed for a proposed 40-inch horizontal telescope which was designed to get star light into the spectrograph with just two reflections and, hence, minimal loss of light. Near the end of the Apollo Moon landings, research and development funds were becoming scarce and the horizontal telescope was never built. Kitt Peak has a similar horizontal telescope to feed the large and very expensive spectrograph of their 84-inch telescope when that telescope is working on other science. In the late 1980s and early 1990s, the coudé focus at the 24-inch telescope was used by Dr. Jeff Kuhn to feed a compact instrument used for solar physics problems.

By late 1970 or early 1971 a Raytheon computer was purchased for the observatory. Although the computer was very limited by today's standards (initially, it had only 16 kilobytes of memory), the plan was to develop an automated photometry system. In the second year of the telescope, a single-channel photoelectric photometer was completed and put into use quickly. Steve Hill, by then on the faculty, proposed a research topic suggested by a paper written by Professor J.D. Fernie, at the University of Toronto. Dr. Fernie was investigating the boundaries of the Cepheid instability strip and studying the possible existence of non-Cepheid stars within it. The instability strip is a region of the HR diagram that was believed to be inhabited only by pulsating stars, but was that true?

We initiated this research project and carried out observations for several years. The question, as we thought, might be one of photometric calibration, but lots of data were required. The MSU photometer we used was based upon the 1P21 phototube and employed the standard Johnson UBV photometric system.

Even in the 1970s, the night sky near the observatory was becoming brighter with expanding development to the north and east, making photography more difficult. Photoelectric photometry quickly supplanted film photography for nearly all the observational work being done. Years later, CCD cameras replaced both traditional photographic cameras and photometers. However, it was the presence of the Raytheon computer that made a huge impact on many of us, for Steve Hill and me, especially.

This was an era when nearly everyone on campus carried boxes of punched cards to the university computer center, submitted card decks to an operator, and waited a day or more to retrieve the printed results. Time-sharing with remote terminals scattered around campus was just beginning. By contrast, since the observatory possessed its own computer, we could work in near real-time, and we did not need to pay the hefty fees charged for using the campus mainframe computer. This was a unique opportunity and an incredible advance in efficiency for us. We had the vision of not only using that computer for research computations, but also using it to automatically collect data and even control the telescope and its instruments.

That Raytheon computer was of special interest to me for several reasons. I had a long-standing interest in positional astronomy (which required computers for long and tedious calculations) and, by that time, I was doing calculations of planetary motions and visibility for the *Abrams Planetarium Sky Calendars*. Steve Hill was using the computer extensively for spherical harmonics computations of pulsating variable stars -- following shock waves as they propagated through the stellar atmospheres. Steve suggested I begin by enrolling in upper level computer science courses, which I quickly did, after which I was to work with him to support the observatory's computer and to develop more powerful software for it.

I had a need for a computer program (or series of programs) to help relieve the burden of calculations and data reductions for the photometry studies. We created those as I developed more programs to aid in the *Sky Calendar* preparations. The computer was expanded to 64 kilobytes of memory and finally its paper tape was replaced by a hard disk drive and later, magnetic tapes. With a tremendous effort on Steve Hill's part, and with our experience in doing photoelectric photometry, we eventually developed an automated photometry system that collected data and controlled the positioning of the telescope and the dome. The Raytheon had a priority driven, multi-tasking operating system (even in a mere 64 kilobytes of memory!), so we could actually do partial data reductions while the computer was moving the telescope to the next star.

Going through the computer courses, one class I opted to take was Cobol programming, just in case I might need to do some business-oriented computer programming. As I had done very well in that class, near the end the instructor told me that should I ever need a job, just call. The outcome was that I was hired by the State of Michigan, eventually transitioning to the State Department of Transportation, where I specialized in engineering and mathematical software, computer graphics, and, later, joined their research program. My experiences at the observatory proved very valuable to me in these positions.

I recently retired after nearly 43 years there. The education I received working with Steve Hill, Tom Stoeckley, and others gave me unique opportunities and it prepared me well for a long career in scientific programming. My MSU experiences created other opportunities for me as well. For example, due to my observing experience and real-time control of telescopes, I was hired, on a year leave from Transportation, to work for the Kitt Peak National Observatory.

Throughout the years a number of undergraduate and graduate students successfully passed through the astronomy program. It has been a pleasure to get to know many of them. One of those talented students was Russell Carroll, with whom I worked later at Kitt Peak on an observing run.

There were also a number of fascinating faculty members, some only briefly at MSU on short-term appointments, and with some of whom I became friends. Allan Saaf, who taught rigorous courses in Galactic Structure and Celestial Mechanics, became an inspiring and treasured friend. Jack Hills shared an interest in computing, in celestial mechanics, and orbits. Jack Sulentic was at MSU briefly, but he and his wife shared with me a love of classical music. I developed computer graphics software for Bob Stein and Jack McConnell. I had a great deal of joy working on a galaxy photometry project as an independent study with Herb Rood; many years later, I helped with a bit more of that work at the McDonald Observatory in Texas. While at MSU, Jeff Kuhn was a most gifted observer, theoretician, and instrument designer and with him I had the rare opportunity of observing with the largest telescopes on Mauna Kea in Hawaii. For Bob Victor, I provided software, calculations, illustrations, and star charts for the *Abrams Planetarium Sky Calendars*, always enjoyable and gratifying work. My MSU and Kitt Peak experiences prepared me, years later, for observing and studying pulsating white dwarf stars at the McDonald Observatory. I owe a great deal to Tom Stoeckley and Steve Hill, for their many contributions; they both deeply enriched my life. Then there is the author of this book, Horace Smith, with whom I share too many interests to list here.

*The 24-inch telescope was used to take this photograph of the double cluster h and chi Persei (NGC 869 and NGC 884).*

# VI. The 1980s: A Near-Death Experience and a Rebirth with Comet Halley

As the 1980s dawned, the state of Michigan, and consequently MSU, were in financial trouble. The inflation and economic stagnation of the late 1970s were followed by a recession in the early 1980s. Michigan's unemployment rate climbed from 7.2% at the start of 1979 to 13.5% in August, 1980. The unemployment rate would peak at 16.5% in late 1982 and would not fall below 10% until 1985. Severe budget cuts hit astronomy particularly hard.

The Astronomy and Astrophysics Department, as it was then named, was faced with closure. With support from the Physics faculty, it was eventually decided to merge the department with Physics, so that, as of mid-1981, there was once more a Department of Physics and Astronomy. However, some of the astronomy faculty left the university and the degree to which astronomy would be supported in the future remained in doubt. The campus observatory shut its doors on July 10, 1981 (Stoeckley 1982). That was the situation when I arrived on the MSU campus in September, 1981. Astronomical sightseeing was once more limited to the planetarium, the naked eye, or small portable telescopes.

Was that the end of the second campus observatory, after little more than a decade of existence? The existence of this book of course discloses that the answer is no. However, the observatory stayed closed for almost five years.

By 1985, economic conditions were looking brighter and the decision had been made to hire additional tenure track faculty in astronomy. Moreover, with the already widely publicized approach of Comet Halley in 1986, interest in astronomy was on the rise. There were finally both motivation and resources to reopen the observatory and Chairperson Jack Bass, together with departmental administrator Marc Conlin[20], acted to make that possible. The 24-inch telescope saw its second "first light", as it were, on the evening of January 21, 1986, when the author (then an assistant professor), visiting professor D. Jack McConnell, and Robert D. Miller found the telescope dusty but operable.

The telescope was open again, though operating on a shoestring budget with no assigned technician and

---

[20] Marc Conlin and Janice Ridenour would for many years be involved with administrative and business tasks connected with the observatory. Staff of the departmental Machine Shop, for much of this time led by Thomas Palazzolo, and including instrument makers Thomas Hudson and Jim Muns, completed vital work at the observatory. Departmental computing support personnel, particularly George Perkins, would also frequently be involved with observatory projects.

without the air conditioning system that had caused so many headaches. Moreover, it was public outreach rather than research that dominated at first. The 1986 apparition of Comet Halley was not a favorable one, particularly from so northerly a locale as Michigan. However, there is no comet more famous and public interest in the comet was great as 1986 began, bolstered by the Giotto and Vega missions to the comet's vicinity.

Comet Halley, sporting a 5-degree tail, was visible to the naked eye low in the southeast on mornings in late March. However, when the comet became visible in the evening in mid-April, it was very low in the south and its brightness had waned. Nonetheless, there was tremendous interest in the comet and a series of public nights attracted crowds of several hundred people each.

Besides myself, these nights were organized and run with the assistance of students, members of the Lansing area amateur astronomy club, and, especially, director David Batch and the staff of Abrams Planetarium. On the best attended of these nights, the line to see the fuzzy coma of the comet through the 24-inch extended out the door of the observatory and down the driveway to Forest Road. On some nights, Comet Halley had set before the last in line reached the 24-inch and later visitors had to be satisfied with a view of Saturn instead. Most, however, at least got a view of Halley through small telescopes set up on the observatory grounds if not through the 24-inch telescope.

All in all, although this apparition of Comet Halley did not rival many previous ones in terms of spectacle, it did make for a busy reawakening of the observatory. By the last of the Comet Halley viewing nights in early May, several thousand members of the public had seen the comet through the 24-inch telescope or through one of the smaller telescopes set up outside the observatory.

By the observatory's 50th anniversary in 2020, comet Halley will be creeping toward its aphelion in late 2023. Who knows? Perhaps the observatory will still be around when Comet Halley has its next perihelion in 2061. I do not know how dark the skies above the campus will be in that year, but the comet's appearance that summer is expected to be superior to that of 1986. I hope that many children who saw Comet Halley at the observatory in 1986 will be around to see the 2061 return in their old age.

With the departure of Halley, the observatory once more reverted to a regular series of public nights, gradually settling into a pattern that would hold for the next few decades. Barring some unusual astronomical event, open house nights were usually held one Friday and Saturday night a month, around first quarter moon, from April until October or early November. Public observing nights were run by the Physics and Astronomy Department and Abrams Planetarium, with help from Lansing area amateur astronomers. Those nights were consistently popular, but rarely would attendance rival those nights with Comet Halley.

It might be remarked that some of the amateur astronomers who helped on public viewing nights were especially dedicated. As an example of such a person, I might mention Kimball (Kim) Dyer (1926 – 2005). Although his eyesight failed as he grew older, so that he was nearly blind in his later years and had to catch a lift to the observatory from his home in the Detroit area, Kim remained a regular figure at open house nights, chatting with visitors as they waited to climb the stairs to the dome. There were many others, unnamed here, who gave freely of their time and expertise.

Professors Stoeckley and Hill had departed MSU before the observatory reopened. Professor Linnell was still at MSU but, although he provided information on technical aspects of the observatory, his research interests had turned in other directions. Thus, the faculty members who had been most involved with the observatory in the 1970s did not participate in its scientific reawakening. Through the 1980s, operations at the observatory were mainly directed by the author and Professor Jeffrey R. Kuhn, who joined the department in September, 1986.

Research at the reopened observatory would be shaped not only by alterations in the composition of the faculty, but also by technological advances in astronomy. Foremost among these was the development of charge-coupled device (CCD) cameras. Faculty member Edwin Din Loh (1948-2019) was a pioneer in the use of CCDs in astronomy before he arrived at MSU (Hearnshaw 1996).

Charge-coupled cameras had a significant advantage over single-channel photoelectric photometry for the MSU observatory. With the photoelectric photometer used in the 1970s, a target star, a comparison star, a check star, and the background sky had to be observed sequentially. Sky conditions could change during the course of this procedure. Not only could a CCD observe more than one star at a time, but its target stars, comparison stars, and the surrounding night sky could be recorded simultaneously. Truly photometric skies are rare in Michigan, but differential photometry can be usefully obtained with a CCD on nights which might be doubtful for single-channel photometry.

Professor Kuhn used some of his startup funding to purchase the first CCD for use at the observatory, a Photometrics, Ltd, Slow Scan camera with a Ford Aerospace/SAIC 1024 x 1024 pixel CCD chip. When used on the 24-inch telescope at the f/8 focus, this liquid nitrogen-cooled camera had a field of view of about 11 x 11 arcminutes. Of course, observing in a heated room with the aid of a computer was much appreciated by both faculty and students during the winter months.

The CCD camera went into operation in 1987 and, after that, direct photography and single-channel photoelectric photometry would rarely be employed. In fact, the photoelectric photometer built in the 1970s would never be used after the telescope reopened in 1986. Nor would students continue to learn a mnemonic to bring to mind the Kodak spectroscopic series photographic emulsions in order from bluest to reddest sensitivity. The one I had been taught and passed along to my students was "Oh John, George doesn't eat flannel underwear nor milk zebras" (OJGDEFUNMZ).[21]

By the 1980s, personal computers had arrived on the scene, and Kuhn, with advice from Professor Linnell, arranged to operate the 24-inch telescope from a small PC/AT computer. The original Raytheon computer was thus bypassed, although the interface of the old system was still employed to communicate with the telescope.

This might be a good spot to note that the aluminum coatings of telescope mirrors do not last forever. Every-so-often the mirrors need to be removed from the telescope and taken to an off-campus facility, where (for a fee) the old coatings are stripped and a new aluminum coating deposited in a vacuum chamber. This has been done several times during the lifetime of the observatory, despite the tendency of those in charge of the observatory to delay the process to avoid the work

---

[21]  The occasion of a visit by Professor Wayne Osborn of Central Michigan University and his Astronomical Techniques class in December, 1988, may have been one of the last times that photographic plates were taken with the telescope.

involved and the interruption to observing. Eventually, the mirror coating becomes sufficiently poor that procrastination is overcome.

Professor Kuhn's interests extended to the Sun and instrumentation for solar research. A heliostat was mounted on the observatory roof that could send light from the sun into the room below. That, and the 24-inch telescope (with most of its aperture blocked to light), were used by Kuhn and graduate student Haosheng Lin to build and test instruments for a high precision photometric solar telescope (Lin and Kuhn 1992). Lin would be awarded his PhD in 1992.

*The primary mirror of the 24-inch telescope, freshly re-aluminized in the spring of 2017. Photograph courtesy of Dustin Scriven.*

# VII. The 1990s: Pulsating Stars, Shoemaker-Levy 9, Bright Comets, and Increasing Light Pollution

The 1990s saw increased research productivity at the observatory and also a number of rare and spectacular celestial sights. Perhaps the most remarkable of these were the collisions of the tidally disrupted fragments of Comet Shoemaker-Levy 9 with Jupiter in July, 1994. However, the observatory skies were also graced by a nearly annular solar eclipse, by bright northern light displays, by two splendid naked eye comets, and by beautiful volcanic sunsets.

The catastrophic eruption of the Mount Pinatubo volcano in June, 1991, would lead to sunset displays reminiscent of those seen after the 1883 eruption of Krakatau[22]. During the late summer and autumn of 1991 and 1992, observers were sometimes distracted from

---

[22] For a discussion of volcanic sunsets, see Meinel and Meinel (1983). Sunset photographs in that book that were taken following the 1963 eruption of the Agung volcano are similar to the Pinatubo sunsets seen in 1991 and 1992. Bishop's rings around the sun, also of volcanic origin, were commonly visible at this time. The aerosol glow after sunset often appeared stronger in 1992 than in the year of the eruption, with that of September 22, 1992, being perhaps the most striking. The volcanic twilight glows weakened in 1993 and were mostly gone by 1994.

setting up for the night's work by spectacularly colored volcanic sunsets. Interestingly, more than a hundred years before, L. G. Carpenter, whom we have already met in connection with the first campus observatory, contributed a note to *Nature* concerning volcanic sunsets (Carpenter, L. G. 1884).

The path of an annular solar eclipse crossed Michigan on May 10, 1994. The eclipse was not quite annular from the location of the observatory, with the northern limit of annularity being a few miles to the south.

*The solar eclipse of May 10, 1994, as seen from the northern limit of the track from which the eclipse could be seen to be annular. Photograph courtesy of Nancy Silbermann.*

Later that same year, an even rarer event occurred. Well before the impact of the first tidally disrupted fragment of Comet Shoemaker-Levy 9 with Jupiter on July 16, 1994, it was known that the collisions would occur. It was not known, however, whether those collisions would leave much of a mark on Jupiter.

A large crowd gathered at the observatory on the evening of July 16, but clouds interfered before the spot of the first impact rotated into view. However, as the collisions continued day by day, their obvious effect upon the appearance of Jupiter soon became evident. I quote from my observing journal for the evening of July 19, 1994: *"When I arrived at the observatory on this rather warm evening, conditions were so cloudy that I despaired of seeing anything. However, the clouds thinned to a haze and by 9:25 pm [EDT] we had the telescope pointed at Jupiter. I was the first to look in the 6-inch and when I did [the others in the dome] heard an excited 'Wow!'"*.

Not very elegant, perhaps, but I was amazed by two large dark splotches, remnants of the Shoemaker-Levy 9 fragment G and L collisions, sprawling across the southern portion of Jupiter. For a time, the impact spots were the darkest and most prominent features on the planet, more evident even than the Great Red Spot or the shadows cast by eclipses of the Galilean moons. Eventually, the spots faded, but they certainly lingered in the memories of those who saw them. Years later, when asked what was the most remarkable sight that I had ever seen with the observatory telescopes, I would reply that it was Jupiter after the Shoemaker-Levy 9 impacts.

By the 1990s, no really bright comet had appeared over the observatory since Comet West in 1976. Then, in 1996 and 1997, two appeared in successive years: Comet Hyakutake followed by Comet Hale-Bopp.

We had not much warning of Comet Hyakutake (C/1996 B2), which was discovered at the end of January, 1996, and which passed 15 million kilometers from Earth in late March, a tenth the distance to the Sun. It developed a spectacularly long tail at that time and, because of its relative nearness, it moved quickly through the night sky. At its peak, it was a fine spectacle for public viewing nights, even though the full length of its tail could not be seen with the light pollution near the observatory.

In contrast, there was plenty of notice for Comet Hale-Bopp (C/1995 O1). It was discovered on July 23, 1995, before Comet Hyakutake, and there were soon predictions that it might become a splendid naked eye sight in the spring of 1997. As was not true for Comet Kohoutek, those predictions proved correct for Comet Hale-Bopp. As a consequence of the long lead time, and the comet's long period of naked eye visibility, Comet Hale-Bopp attracted much public attention.

Comet Hale-Bopp was a public viewing night target in observatory telescopes by the summer of 1996, reaching naked eye visibility from the observatory grounds toward the latter part of the year. Cloudy winter skies hindered observations early in 1997, but with clearer skies the comet was a beautiful sight on many mornings and evenings in March, April, and early May.

*Comet Hale-Bopp photographed on April 9, 1997. Author's photograph.*

The observatory, with Lansing to the west, was not an ideal location for seeing Hale-Bopp when it was visible in the evening. Nonetheless, many visitors showed up at the observatory hoping to see the comet through a telescope, even on nights when no public viewing had been announced. For a while, a student with an 8-inch telescope was stationed outside the observatory each clear night to show the comet to the public, so that other students inside were free to carry out undisturbed their observations with the 24-inch telescope.

When the observatory reopened in 1986, the undergraduate observatory practice course, AST 437, which had existed in the 1970s, was not resumed. For a few years, observing was made a part of the upper level AST 327, Practical Astronomy, course. However, that course vanished when the university switched from the quarter to the semester system in the fall of 1992. Observing then became part of the spring semester sophomore level AST 202 course aimed at astrophysics and physics majors. When undergraduate astronomy courses were once more rearranged in the early 2000s, observing was for a few years put in the stand-alone AST 312 laboratory course. That course, too, would eventually disappear as observing was subsumed into AST 208, Planets and Telescopes, in 2005. It remains there as of this writing.

Since the switch to the semester system, the observing courses were all taught during the spring semester, which of course really starts in wintery January. Students and teachers often lamented poor observing conditions during the first half of the semester, but somehow it never proved possible to change the course to the usually more clement fall semester. Nonetheless, the number of students taking those courses followed an upward trend from the 1990s into the 2000s. Of course, once they had their training, undergraduates used the observatory for senior thesis projects and other research throughout the year.

When the Physics and Astronomy Department moved to the new Biomedical and Physical Sciences building in 2002, rooftop piers were provided for mounting portable 8-inch telescopes. However, most observing remained centered at the observatory.

CCD imaging cameras remained the main instruments used on the 24-inch telescope during the 1990s, although, as we saw in the previous chapter, Professor Kuhn also used the observatory for developing solar instrumentation. When the Photometrics CCD was no longer used at the observatory, CCD cameras were purchased that used various thermoelectric cooling systems rather than liquid nitrogen. Though such cooling methods did not cool the detector to as low a temperature as did liquid nitrogen, they were easier for students to use on short notice, and suitable for the observatory's light polluted sky. After some trial and error, the old Raytheon computer room would eventually be chosen as the control room for remotely operating the CCD camera. During this decade the internet also reached the observatory, changing the way that observers approached a night at the telescope. By the end of the decade, carrying tapes of data away at the end of the night was a thing of the past.

Short-period pulsating stars became the chief targets for CCD photometry as the 1990s progressed. By short-period, we mean stars such as delta Scuti stars or RR Lyrae stars, which generally have pulsation periods of less than a day. Nancy Silbermann carried out extensive CCD observations of variable stars in the globular cluster M15, observations which formed a substantial portion of her 1994 PhD dissertation. An observing program was also begun on multimode field RR Lyrae stars, and the number of published papers based upon observations with the 24-inch telescope increased significantly (see Appendix I, Appendix II, and Chapter X). These research programs, often supported in part by the National Science Foundation, involved graduate students, undergraduate astrophysics and physics majors, and summer students in the Research Experiences for Undergraduates (REU) program.

When Professor Kuhn left MSU, first on a temporary leave and then permanently in the late 1990s, I became the faculty member responsible for directing almost all of the research carried out at the observatory. Professor Suzanne Hawley, a new faculty member in 1993, also became involved in observatory operations, but left MSU at the end of the decade. Among her numerous later achievements, Dr. Hawley would be director of the 3.5-m ARC telescope at the Apache Point Observatory.

As the 1990s progressed, research at the observatory was increasingly limited to relatively bright stars because of worsening light pollution. The Pavilion for Agriculture and Livestock Education opened east of the observatory in 1996. Parking areas between the observatory and the Pavilion became a sea of recreational vehicles when weekend shows were underway. Meanwhile, there was increased building throughout the southern part of the campus and the southern portions of East Lansing and Meridian township. The still relatively dark sky of the early 1970s became a thing of the past.

At the start of the 1990s, it was still possible to see stars of magnitude 5.4 with the naked eye at the zenith on a good night. By the 2010s, it was rare that stars fainter than magnitude 4.5 could be seen with the unaided eye. In 2009, tests of the light at the zenith on clear and moonless nights gave a sky brightness in the visual of about 19.3 magnitudes per square arcsecond. For comparison, the corresponding number at a dark location in Michigan's Upper Peninsula is about 21.5, a factor of 8 less light. Away from the zenith, the sky was even brighter, corresponding overall to a Bortle scale sky of 6 even under the most favorable circumstances.[23] Nonetheless, scientifically significant CCD photometry could continue, provided that target stars were carefully selected.

Routine maintenance of the observatory building was carried out, but sometimes things did not quite go as planned. A company hired to repaint the exterior of the dome used a sandblasting technique to remove the old coating of paint. However, a large amount of grit ended up inside the dome! It took more than a bit of effort to remove it and, for years afterward, fine pieces of grit would reappear from hidden crevices when the wind blew strongly[24]. Lesson: even if someone else is doing the work, pay attention to the details.

While research continued at the 24-inch, there was no denying that a relatively small telescope in mid-Michigan could not meet all of the observational needs of the existing faculty, nor would it serve to attract and retain high quality new faculty. Faculty members could and did successfully apply for time on telescopes at Kitt Peak National Observatory, Cerro Tololo Interamerican Observatory, and other facilities. However, as the 1980s

---

[23] John Bortle's scale for judging sky darkness and light pollution is described at http://www.skyandtelescope.com/astronomy-resources/light-pollution-and-astronomy-the-bortle-dark-sky-scale/ .

[24] The grit also caused long-term damage to the motion of the dome shutter.

moved into the 1990s, the Physics and Astronomy Department sought more direct access to a larger optical telescope.

As a stopgap, time was purchased on and instrumentation supplied to the 2.3-m telescope of the Wyoming Infrared Observatory. However, explorations continued concerning MSU participation in other telescope projects, including the Large Binocular Telescope and the Magellan telescopes. In 1996, Michigan State University joined the SOAR (Southern Astrophysical Research Telescope) consortium[25] in its efforts to build a 4-m class telescope at Cerro Pachon, Chile.

Science observations with the 4.1-m SOAR telescope would begin in 2005 and, a few years later, thanks mainly to the efforts of Professor Edwin Loh, MSU would, with help from Brazil and other partners, provide the telescope with the SPARTAN near-infrared camera.

As might be expected, it was the SOAR telescope that occupied most of the time of the observers on the faculty, and much of the funding available to astronomy. In addition, the Joint Institute for Nuclear Astrophysics, an NSF-supported collaboration including MSU, began in 2003 and it became involved with the Sloan Digital

---

[25] The SOAR consortium included NOAO, the University of North Carolina, Brazil, Michigan State University, and Chile.

Sky Survey (Suelter 2008, p. 235-236). The campus 24-inch telescope kept chugging away, as it were, but it would remain very much a matter of secondary attention during these years.

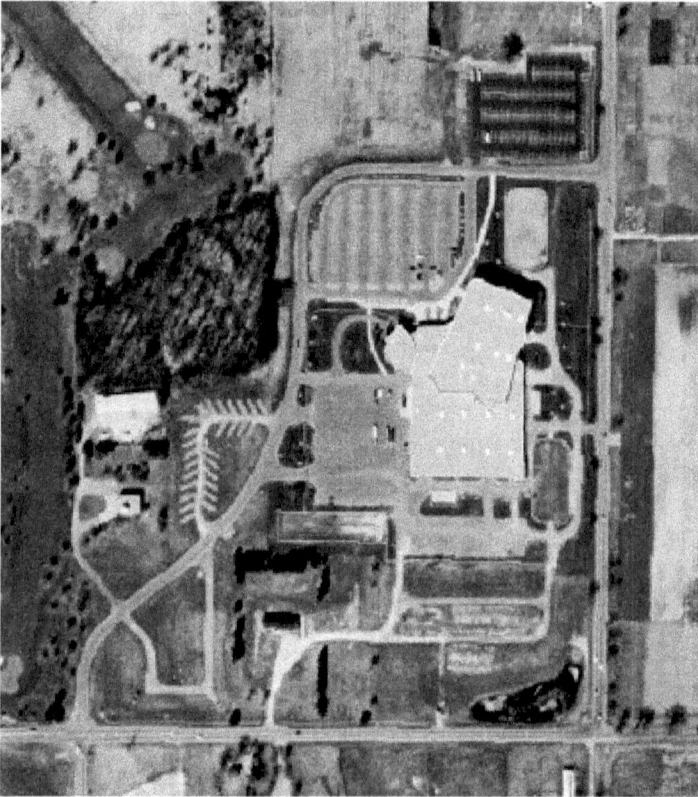

*This 1998 satellite image shows the Pavilion for Agriculture and Livestock Education to the east of the observatory, while Forest Akers Golf Course has expanded in the west. North at top. East at right.*

# VIII. The 2000s: Peculiar Comets, Mars, and a Major Renovation

By the middle of the first decade of the 21st century, the observatory was showing its age. While the telescope optics, mounting, and gearing were still good, other aspects of the observatory were in poorer shape. The dome had rusted in places; there was no modern computer system for operating the telescope; and the positional coordinates for the telescope began to fail in cold weather. Clearly, it was time to seek money for a major renovation.

At this time the National Science Foundation was grappling with how to maintain a national system of telescopes for astronomy when budget constraints threatened continued operation of the older telescopes of the National Optical Astronomy Observatories. The NSF began a new grant program in 2004, the Program for Research and Education with Small Telescopes (PREST). This program would provide modest grants supporting 0.5 – 2.5-m aperture telescopes at universities and other institutions. One motivation for this program was the idea that it would improve the overall national infrastructure in astronomy, with small private telescopes taking up some of the slack from diminished support of small telescopes at Kitt Peak National

Observatory and Cerro Tololo Interamerican Observatory.

PREST would not be a long-lasting program at NSF, but it provided an opportunity for MSU. The MSU observatory by then had an established track record for student involvement in research, and PREST's emphasis on student participation in research and training seemed a good match.

With the help of other faculty members, I prepared and submitted in 2004 a PREST proposal for renovation of the observatory. This effort was successful and MSU was awarded a PREST grant of $144,971. The renovation of the telescope, with supplementary funding by MSU, was on. The contract for the major part of the work was eventually awarded to Astronomical Consultants and Equipment, Inc. (ACE) of Tucson, Arizona.

In 2005, Peter Mack of ACE installed a new computer-controlled operating system for the telescope and a new back end to the telescope with filter wheel, to which a CCD or an eyepiece could be attached. The old control console was removed from the dome and, at about this time, the last remnants of the Raytheon computer and the old telescope interface were removed. University funding provided for renewal of the dome, and staff from the departmental machine shop later repaired its worn shutter motion.

One mistake was made at this time. The two drive motors that had turned the dome were replaced by a single motor. It would later turn out that this could occasionally result in the dome being stuck in position, while the single motor futilely spun its driving wheel and a smell of burning rubber filled the dome. Several years later a second drive motor and wheel would once more be added, resolving the problem.

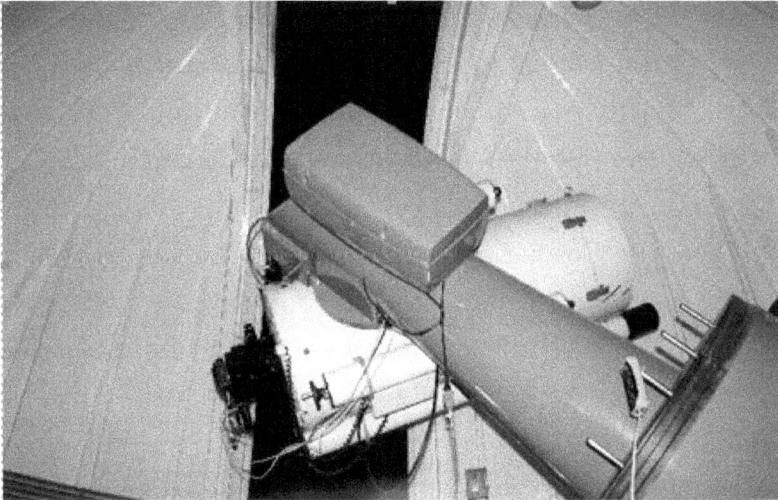

*An Apogee CCD is mounted on the refurbished Boller and Chivens telescope. Author's photograph.*

These renovations did not make the telescope into a completely robotic instrument. Someone still had to be at the observatory to take the cover off the telescope and to switch instruments on. Nor was there any weather sensor to detect clouds or rain. However, after the renovations, the observatory was much closer to being adapted for remote use, should that be deemed desirable in the future. So far, it has been thought that actually being at the observatory has benefits for student education, so that there has been no attempt to raise the extra funds to go entirely robotic.

The public outreach highlight of the early 2000s was the close approach of Mars in 2003. During the 2003 opposition Mars was only a tad closer to the earth than at other favorable oppositions. Nonetheless, the fact that on August 27 Mars would technically be closer to the Earth than it had been for thousands of years caught the public attention. Attendance at the September 5 and 6 public viewing nights rivaled those for Halley's Comet, although even at a close approach Mars is a more difficult visual target than Jupiter or Saturn. John French of Abrams Planetarium estimated that 2,000 visitors were able to see Mars during the two nights.

Transits of Venus in 2004 and 2012 also attracted considerable interest, but most of the public observing for those events was held on the parking ramp east of Abrams Planetarium rather than at the observatory.[26] The spectacular Leonid meteor display of 2001 was

---

[26] I have not discovered whether the transit of Venus on December 6, 1882 was seen from the first campus observatory. Clouds hindered observing in parts of Michigan.

fogged out at the observatory, leaving prospective observers dashing for their cars to find a clearer location (some were successful).

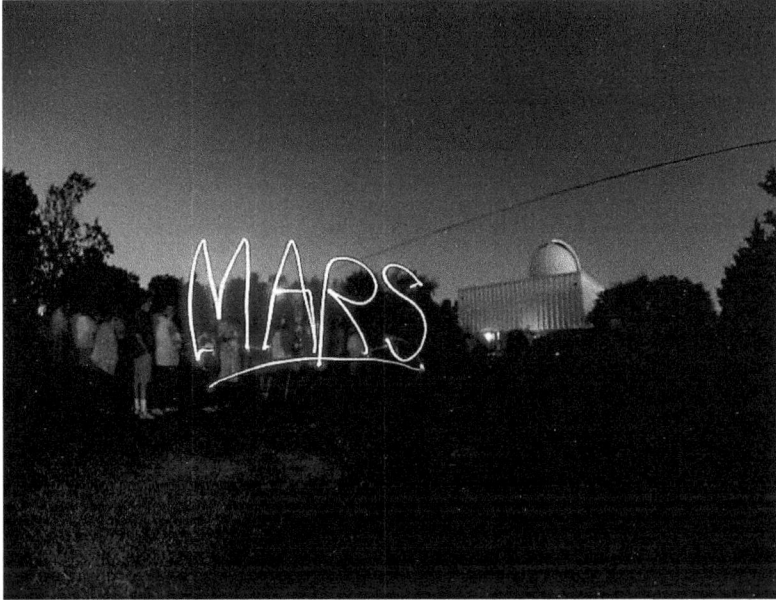

*A moving flashlight spells out the target during this September, 2003, open house night. Photograph courtesy of John French.*

The year 2009 saw one event reminiscent of the Shoemaker-Levy 9 impacts of 1994. Again, a dark spot appeared on Jupiter that was attributed to an impact event. This time, however, the spot was much smaller and less durable than the large impact spots of 1994.

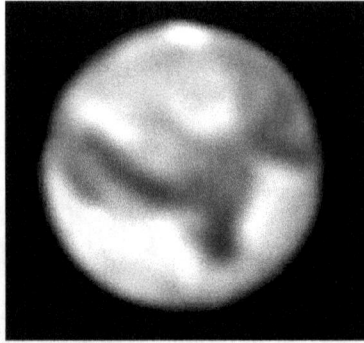

*In 2003 the CCD used at the 24-inch telescope was chosen for stellar research and was far from ideally suited for planetary imaging. Nonetheless, some images of Mars were taken for fun. This is a near-infrared Cousins I-band image from September 8, showing the polar cap and Syrtis Major. South at top.*

Fragmenting comet 73P Schwassmann–Wachmann 3 passed relatively near the Earth in 2006, allowing some nice images of the phenomena associated with its break-up. Fragment C of this comet passed M57 (the Ring Nebula) on the night of May 7/8, while Fragment B suffered further fragmentation with an associated burst in brightness. Though the research program of the observatory had little to do with comets, the CCD was used to record these events.

*The 2009 impact spot, a little hard to see in this image from the 24-inch telescope, was of course much more clearly seen in Hubble Space Telescope images. North at top.*

Just before October 23, 2007, Comet 17P/Holmes was far below naked eye visibility, fainter than apparent magnitude 14. That changed quickly. An unexpected outburst sent it surging to naked eye visibility on October 24. Before the end of that day the comet shone like a 2nd magnitude star in the constellation Perseus.

Successive images of 17P/Holmes with the CCD camera on the 24-inch telescope showed a rapid expansion of the comet's coma, which soon exceeded the camera's 10 x 10 arcminute field of view. It would turn out that this was not the only outburst of Comet Holmes. An outburst in 1892 had also lifted the comet to naked eye visibility, but the 2007 outburst was greater still.

Between 2000 and 2009, the research focus of the observatory remained variable stars. I and my students used the telescope and successive CCD cameras to observe pulsating stars in the field and in globular clusters. The program of photometry of multimode RR Lyrae stars that had been begun during the 1990s was continued, with the bright field stars XZ Cygni and RR Lyrae itself being particular targets.

B, V, and Cousins I-band imaging of globular clusters was begun, with particular emphasis on M3, M5, and M13. The purpose of this program was to obtain new light curves for type II Cepheids in those clusters and to study the long-term period changes of those stars. Cepheids, like RR Lyrae stars, are pulsating stars, but they have longer periods, often between one and thirty days. MSU photometry of these Cepheids would eventually be combined with observations from telescopes at other institutions, and with observations from the literature, providing time-spans of observation in some cases longer than a century. In general, increased collaborations with other observatories and institutions would distinguish the research done during this decade.

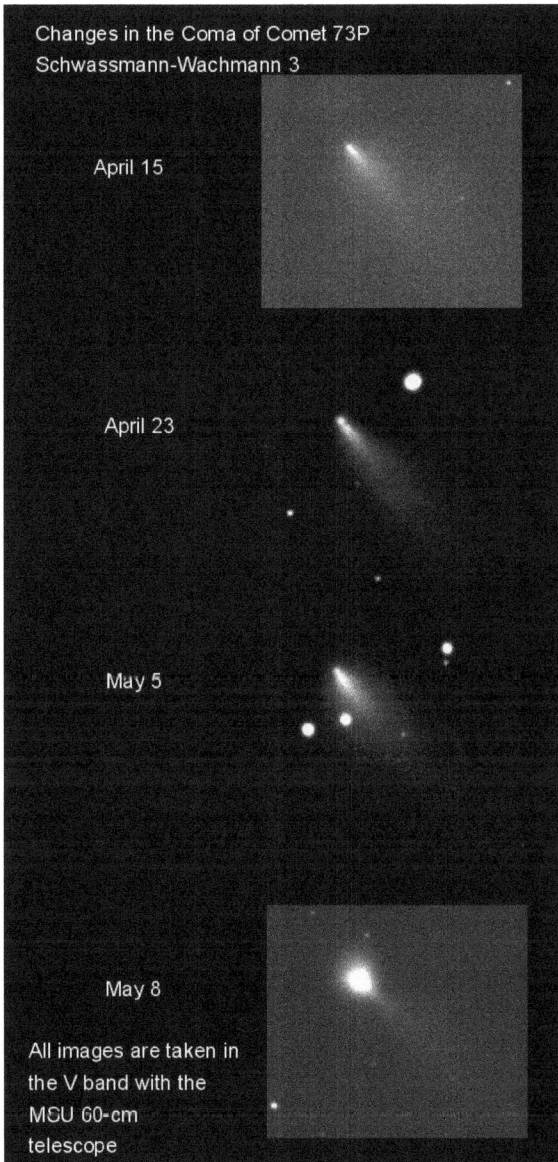

Changes in the Coma of Comet 73P
Schwassmann-Wachmann 3

April 15

April 23

May 5

May 8

All images are taken in
the V band with the
MSU 60-cm
telescope

*Fragmentation and outburst of Fragment B of Comet
73P Schwassmann-Wachmann 3 in 2006.*

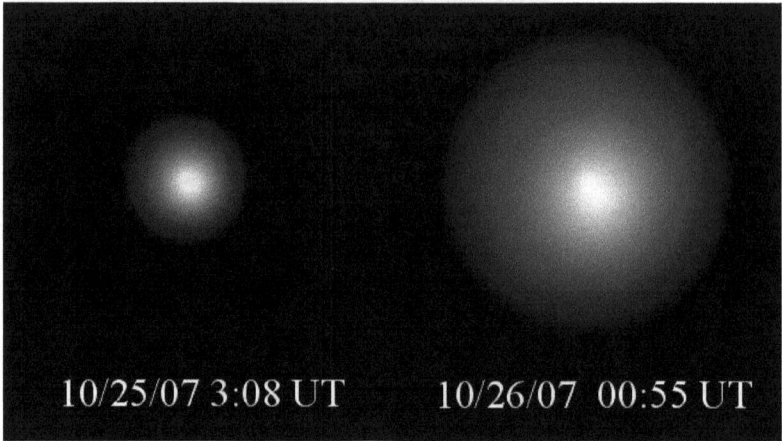

10/25/07 3:08 UT          10/26/07  00:55 UT

*One day's change in the apparent size of Comet 17P/Holmes, as recorded in the V band with the 24-inch telescope and CCD camera. The second image nearly fills the 10 x 10 arcmin field of view.*

# IX. The 2010s: A Change in Direction and Director

As the second decade of the 21st century arrived, my time on the MSU faculty was drawing to a close. What would become of the telescope after my retirement in 2013? The 24-inch telescope was not suited to the research endeavors of existing departmental faculty and their students, whose observational research was more likely to require the much larger SOAR telescope or telescopes at other institutions.

Fortunately for the continued operation of the observatory, a new faculty member, Laura B. Chomiuk, was hired who, among her other interests, wished to maintain student research and outreach at the observatory. She first came to MSU as a Jansky Fellow but was appointed a professor in 2013.

When Professor Chomiuk assumed direction of the observatory, she put some of her startup funds into a return visit to the observatory by Peter Mack of Astronomical Consultants and Equipment, Inc., resulting in installation of a second dome motor and updating of the telescope control software. A CCD for guiding the main telescope was added to the 6-inch refractor at this time.

Though a few experiments with a small spectrograph had been made in the 1990s, the 24-inch was almost always used for direct imaging in that and succeeding decades. However, the possibility of spectroscopy returned to the observatory in 2018, when a light-weight Shelyak Lhires III spectrograph was obtained that could be mounted at the Ritchey-Chretien f/8 focus.

At about this time there were other changes made at the observatory that were less directly linked to the scientific uses of the telescope. The observatory plumbing fixtures and water fountains were replaced early in 2016, making the structure seem less a product of the 1960s, and giving visitors better tasting water. The heating system was updated. Following a university-wide trend, entry to the building was changed from the traditional key to a keycard system. From the autumn of 2015 into 2016 the observatory driveway was long disrupted as a new underground electricity conduit for the campus was constructed, but the driveway ended up much the same afterward, aside from some renewed paving.

As the decade drew toward its close, there was, alas, no let-up in development that threatened additional light pollution. McLaren Hospital announced that it would move its two south Lansing facilities to a new nine-story building to be constructed on one-time farmland between Collins Road and US 127, south of

Forest Road and to the southwest of the observatory. Construction of the new facility is underway as this is written.

Before Professor Chomiuk took over the observatory, the research program remained much as it had been in the previous decade. However, there was increased attention to extragalactic supernovae, due to the appearance of bright supernovae in the nearby spiral galaxies M51 and M101 in 2011. Type Ia supernova 2011fe in M101 climbed to V = 9.9 at its peak brightness, making it one of the apparently brightest northern supernovae in recent decades. Photometry from the Rochester Institute of Technology was combined with that from the campus observatory to produce a light curve covering the first half year after its outburst (Richmond and Smith 2012). Observations of supernovae were also contributed to the database of the American Association of Variable Star Observers.

Once Professor Chomiuk assumed control of the observatory, its research program shifted. Since 2014, she has engaged a large number of undergraduate students in research at the observatory (~20 at a given time) through the MSU Observatory Research Program (MORP). The central focus of this research is photometric monitoring of cataclysmic variable stars in collaboration with the Center for Backyard Astrophysics (directed by Joe Patterson of Columbia University). Student observations have been used in several senior theses and in publications (Littlefield 2016; Patterson et al. 2017a). MSU postdocs Kwan-Lok "Ray" Li and Evangelia Tremou also monitored a binary millisecond pulsar with the observatory and measured its orbital period (Li et al. 2016).

*Supernova 2011dh is identified on this image of M51 taken with the CCD camera on the 24-inch telescope on June 30, 2011.*

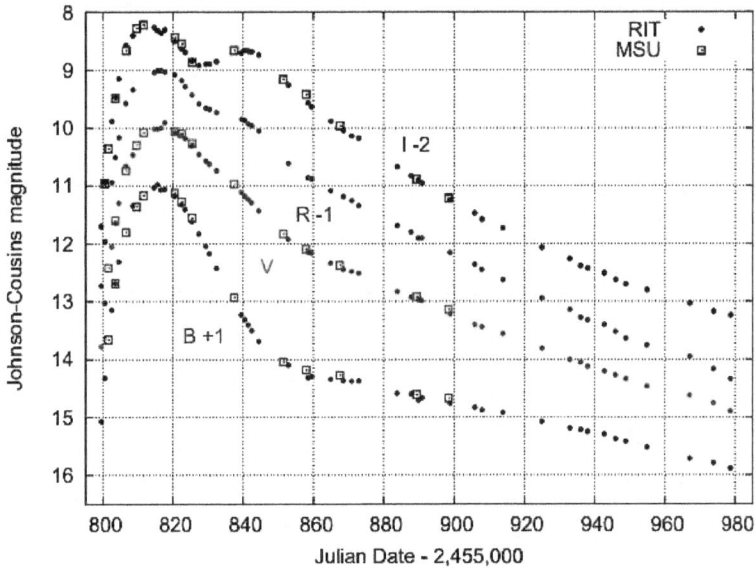

*The light curves of supernova 2011fe in M101 in the Johnson-Cousins BVRI system, from Richmond and Smith (2012).*

In 2018, in response to strong student interest in exoplanets, the observatory joined the Follow-Up Network for the Kilodegree Extremely Little Telescope (KELT FUN; Collins et al. 2018). The task of the MSU observers is to help vet candidate exoplanets through photometric monitoring of (suspected) transits.

Professor Chomiuk continued and expanded public outreach efforts at the observatory, with the aid of Shannon Schmoll, who succeeded David Batch as director of Abrams Planetarium in 2014. Chomiuk and her students began to install more formal educational exhibits at the observatory. 2018 saw the installation of illuminated astronomical images along the hallway walls.

The observatory, together with Abrams Planetarium, helped found Statewide Astronomy Night, which is held each April in conjunction with the MSU Science Festival. It is often attended by 200 - 300 members of the public. The observatory also has an annual open house in October for International Observe the Moon Night. Those two special open house nights are accompanied by various hands-on activities as well as the usual observing.

News of events at the observatory, such as open house nights, had originally been circulated by word of mouth, flyers, and the occasional newspaper or radio announcement. In the 1990s, these were supplemented by postings on the departmental and planetarium websites. With the growth of social media in the 2010s, the campus observatory received its own Facebook and Twitter accounts, mainly orchestrated by undergraduate students. As of this writing, the observatory had some 1900 followers on Facebook and 280 on Twitter.

The track of totality for the August 21, 2017, eclipse of the Sun passed well south of Michigan. Nonetheless, a very large crowd turned out at the observatory on eclipse day, enjoying the partial eclipse. With many faculty (including your author, by then of course an emeritus professor) and planetarium staff skipping town to see totality, it remained for students,

*Professor Chomiuk encourages students in the observatory control room, once the Raytheon room. Observatory collection.*

led by Huei Sears, and postdocs, aided by Professors Steve Zepf and Edward Brown, to take charge of the successful event. The close Mars opposition of 2018 also attracted many visitors to open houses, although a planet-wide dust storm dimmed Mars's features when it was nearest Earth.

In 2018, it was decided that an observatory logo should be created to accompany other aspects of its new social media presence. With Professor Chomiuk's encouragement, student observers Ryan Smith, Daniel

Coulter, Dylan Mankel, and Huei Sears devised one featuring MSU's traditional green and white colors.

*The logo for the campus observatory designed by students in the summer of 2018. Imagine the gray as an MSU green.*

*Aerial view of the observatory in the summer of 2019. Looking to the southeast. Courtesy of Robert D. Miller.*

# X. A Look Back and A Look Ahead

This history is being completed in 2019, just ahead of the 50th anniversary of first light at the second campus observatory. In the preceding chapters I've reviewed many aspects of those fifty years, but it is convenient in this chapter to summarize some of them. Numbers can help here, so let's begin with a healthy dose of tabulations.

It is difficult to reliably estimate how many members of the public have come to the second observatory during five decades of open house nights. A number of 50,000 visitors is not too large, although some of them would be repeat customers, as it were. What long-term influence those nights at the telescope may have had on members of the public I cannot say, but I can assert that the memories of those nights have stuck with some visitors for decades. They have told me so.

There is no tally of all of the times that a newspaper, radio, or television reporter came to the observatory or inquired about it. That tended to happen when some astronomical event was in the news, such as a bright comet or an eclipse, but occasionally roving reporters would arrive just seeking something a little novel. With the exception of the 1976 article in the *Detroit Free Press* mentioned in Chapter V, the resultant publicity has been advantageous. A 2009 program on the observatory made by the Meridian Township government access channel, HOMTV was one of the more extensive productions.

Now we turn to the observatory's contributions to science. As listed in Appendix II, by September, 2019, observations made at the observatory had contributed to 38 papers in refereed journals, 12 papers in conference proceedings or unrefereed publications, 25 abstracts or short notices, two MS theses, and two PhD dissertations. The authors of these publications frequently included undergraduate, graduate, and summer students, some of whom were authors of more than a single paper. For many students, observing with the 24-inch marked their first acquaintance with scientific research.

It is noteworthy that many recent papers resulted from collaborations that extended beyond Michigan State University, to other institutions within the United States or in other countries. If we peruse the papers listed in the Appendix, we see that, through the 1990s, observatory publications only rarely involved cross-institutional collaborations. Collaborations with other institutions, with international partners, and with multi-facility research programs became the norm in the decade after 2000. Doubtless, such collaborations were made more possible by the ease of communication and data exchange afforded by the internet. However, the rise of collaborations also signals a change in the type of research open to a modest telescope. In the realm of variable star astronomy, it has become harder for a single, small telescope to match the work done by multi-telescope collaborations.

The subjects of research publications have also changed over time, being strongly driven by the different interests of the faculty directing telescope use. This is illustrated in the charts below, which plot the number of total publications by year and type and the number of refereed papers by year, distinguishing five research areas: (1) instrumentation and equipment, (2) eclipsing binary stars, (3) pulsating stars, (4) cataclysmic variable stars, plus (5) a catchall "other" category. Cataclysmic variables that also eclipse have been kept in the cataclysmic star category. The hiatus in papers resulting from the closure of the observatory from 1981 until 1986 is evident, with the resumption in publications not coming until a few years after the reopening.

We see from the charts that, while research papers have dealt with a number of subject areas, the greatest number of papers have concerned pulsating stars. This reflects my own research interests during the years that I was in charge of the observatory. Undoubtedly, we shall see an upswing in other research areas in the years ahead, in particular perhaps cataclysmic variables and topics connected with the

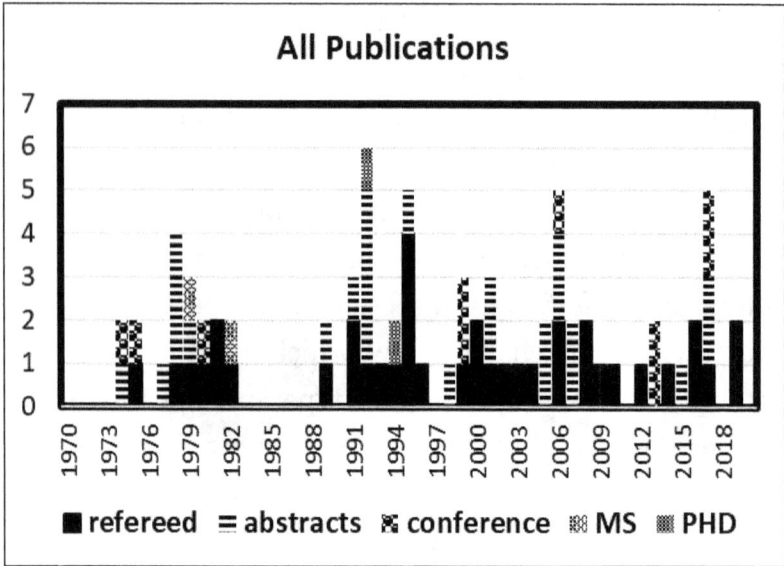

*All Publications, by year, 1970-2019.*

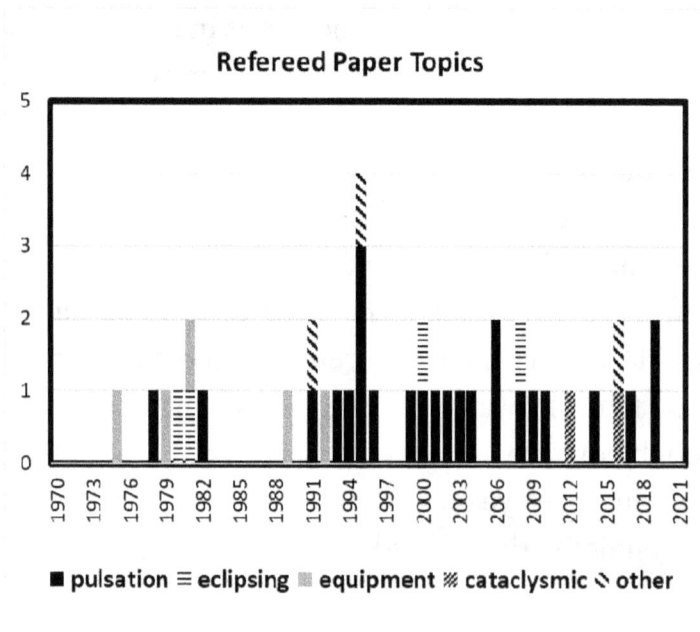

*Refereed Publications, by year, 1970-2019.*

KELT collaboration. While we cannot claim the campus observatory to be one of the nation's chief astronomical research facilities, we can certainly credit it with useful contributions to astronomy during its first half century.

In Chapter IV, we noted that, in 1970, Michigan State University was not alone in operating a midwestern observatory. What is the case five decades later? Many of the larger midwestern observatories operating in the 1960s have either closed or moved away from the Midwest. Not all are gone, however. The Ritter Observatory and the Perkins Observatory still exist as research and educational facilities at their original locations. The Pine Bluff Observatory also exists, but with limited activity. Moreover, remaining older observatories have been joined by new midwestern observatories fulfilling teaching, outreach, and research functions.

These newer observatories have telescopes of modest size, with apertures comparable to, although often a bit smaller than, that of the MSU 24-inch telescope. Calvin Observatory, on the campus of Calvin College in Grand Rapids,[27] and the Brooks Observatory of Central Michigan University are two examples which have origins roughly contemporaneous to that of the second MSU observatory. To the north, Michigan

---

[27] Now also associated with the robotic Calvin-Rehoboth observatory in New Mexico.

Technological University acquired the Amjoch Observatory in 1995. Nor are college observatories the only ones surveying Michigan skies. Michigan amateur astronomers, some involved in research activities, have built a variety of facilities over the past few decades, including the Fox Park Observatory in Potterville, just outside of the Lansing area.

While its main research telescopes are now far from its home state, the University of Michigan has not completely abandoned Michigan as a site for observing. It operates a 16-inch reflector atop Angell Hall in Ann Arbor, and maintains the original Detroit Observatory, but as an historical and educational rather than a research facility.

Nor are newer Midwestern observatories absent outside of Michigan. Built in the 1980s, the campus observatory at Bowling Green State University in northern Ohio is only one of a number of new observatories within the Great Lakes region.

Technology developed since the 1960s has allowed research to continue into the 21$^{st}$ century at these and other small observatories. It is probably safe to say that, without CCDs and similar electronic detectors, little research could now be conducted from midwestern observatories, battling as they do clouds, poor seeing, and light pollution. It is also noteworthy that many of these observatories, including the MSU observatory, emphasize time-series studies of variable or moving astronomical objects.

How long will it be worthwhile for Michigan State University to maintain an establishment such as the campus observatory? That is hard to answer. Developments in technology have changed the research niches open to smaller telescopes. As a research instrument, the 24-inch telescope will in the future need to be competitive in an environment where large CCD surveys increasingly monitor the skies. So far, the observatory has kept a research role by joining larger collaborations and by bringing intensive observations to bear on target stars that have been lightly covered by more sweeping surveys. The time when a 24-inch telescope, in an observing site such as Michigan, can no longer advance the field may come, but not, I think, within the next decade and possibly for much longer.

There may also come a day when the observatory loses its role as an education and outreach instrument. Perhaps there will be a time when, instead of visiting a local observatory, members of the public desirous of seeing astronomical sights don virtual reality helmets to scan the skies as they might appear from pristine observing locations around the world (or in space, for that matter). That does sound rather cool, but I think that something would be lost if the real, local, sky were to be entirely abandoned. In any case, I think that such a time, too, is many years away.

More than anything else, the immediate future of the campus observatory will depend upon MSU faculty, administrators, and students. It will be they who will need to do the work to keep the telescope going, they who will need to develop and carry out the research programs, and they who will need to show the skies to an inquiring public. I wish them well.

*The author at the observatory after a public viewing evening in 2016. Photograph courtesy of Mary Anderson.*

# References

Barnard, E. E., 1889, Publications of the Astronomical Society of the Pacific, **1**, 89.

Beal, W. J., 1915, History of the Michigan Agricultural College and Biographical Sketches of Trustees and Professors, published by the Agricultural College, East Lansing.

Bell, Trudy E., 2002, Journal of the Antique Telescope Society, 23, 9.

Bullis, E.P. and Campbell, L., 1959, Moonwatch Catalogue- September 1958, SAO Special Report Number 20. Part 6.

Bulletin for Visual Observers of Satellites, Smithsonian Astrophysical Observatory, No. 2, October 1956.

Cameron, G. L., 2010, Public Skies: Telescopes and the Popularization of Astronomy in the Twentieth Century, PhD dissertation, Iowa State University.

Carpenter, L. G., 1884, Nature, 30, No. 758, 32.

Carpenter, R. C., 1876a, Report of the Department of Engineering and Mathematics, in the Fourteenth Annual Report of the Secretary of the State Board of Agriculture of the State of Michigan for the Year 1875, W. S. George and Co., p. 58.

Carpenter, R. C., 1876b, Report of the Department of Engineering and Mathematics, in the Fifthteenth Annual Report of the Secretary of the State Board of Agriculture of the State of Michigan for the Year Ending September 30, 1876, W. S. George and Co., pp. 125-126.

Carpenter, R. C., 1878, Report of the Department of Engineering and Mathematics, in the Seventeenth Annual Report of the Secretary of the State Board of Agriculture of the State of Michigan for the Year Ending August 31, 1878, W. S. George and Co., pp. 113-114.

Carpenter, R. C., 1879, Report of the Department of Engineering and Mathematics, in the Eighteenth Annual Report of the Secretary of the State Board of Agriculture of the State of Michigan for the Year Ending August 31, 1879, W. S. George and Co., pp. 41-42.

Carpenter, R. C., 1880, Report of the Department of Engineering and Mathematics, in the Nineteenth Annual Report of the Secretary of the State Board of Agriculture of the State of Michigan for the Year Ending August 31, 1880, W. S. George and Co., pp.58-59.

Carpenter, R. C., 1882, Report of the Department of Mathematics and Civil Engineering, in the First Biennial Report of the Secretary of the State Board of Agriculture of the State of Michigan from September 1, 1880, to September 30, 1882, W. S. George and Co., pp. 154.

Carpenter, R. C., 1884, Report of the Department of Mathematics and Civil Engineering, in the Twenty-Third Annual Report of the Secretary of the State Board of Agriculture of the State of Michigan from October 1, 1883 to September 30, 1884, W. S. George and Co., pp. 50.

Collins, K. A., et al. 2018, Astronomical Journal, **156**, 234.

Foote, J. L., 2005, The Center for Backyard Astrophysics: Theory and Practice, Society for Astronomical Sciences Annual Symposium, **24**, 91.

Hearnshaw, J. B., 1996, The Measurement of Starlight, Cambridge University Press, Cambridge, p. 478.

Hockey, T., 1999, Galileo's Planet: Jupiter Before Photography, Institute of Physics Publishing.

Howard, Neale E., 1958, Handbook for Observing the Satellites, Thomas Y. Crowell Company.

Lin, H. and Kuhn, J. R., 1992, Solar Physics, **141**, 1.

Linnell, A. P., 1972, Bulletin of the AAS, **4**, 291

Linnell, A. P., 1981, Publications of the Astronomical Society of the Pacific, **93**, 661.

Linnell, A.P. Hill, S. J., and Brandt, E. F., 1975, Publications of the Astronomical Society of the Pacific, **87**, 273.

McCray, W. Patrick, 2008, Keep Watching the Skies! The Story of Operation Moonwatch and the Dawn of the Space Age, Princeton University Press.

Meinel, Aden and Meinel, Marjorie, 1983, Sunsets, Twilights, and Evening Skies, (Cambridge University Press).

Musto, D.F., 1967, A survey of the American Observatory Movement, 1800-1850, Vistas in Astronomy, **9**, 87.

Patterson, J., et al., 2017a, Society for Astronomical Sciences, **36**, 1.

Percy, J. R., 1986, The Study of Variable Stars Using Small Telescopes, Cambridge University Press.

Richmond, M. W. and Smith, H.A., 2012, Journal of the American Association of Variable Stars Observers, **40**, 872.

Rogers, J. H., 1995, The Giant Planet Jupiter, Cambridge University Press.

Stephens, D. C. et al., 2016, AAS/Division for Planetary Sciences Meeting Abstracts, **48**, 122.01.

Stoeckley, T. R., 1979, Bulletin of the AAS, **9**, 176.

Stoeckley, T. R., 1978, Bulletin of the AAS, **10**, 241.

Stoeckley, T. R., 1979, Bulletin of the AAS, **11**, 222.

Stoeckley, T. R., 1980, Bulletin of the AAS, **12**, 246.

Stoeckley, T. R., 1981, Bulletin of the AAS, **13**, 256.

Stoeckley, T. R., 1982 Bulletin of the AAS, **14**, 306.

Suelter, C. H., 2008, A History of the College of Natural Science at Michigan State University, 1855 – 2005, (Michigan State University), DVD.

Vedder, H. K., Report of the Department of Civil Engineering, in the Fifty-Fourth Annual Report of the Secretary of the State Board of Agriculture of the State of Michigan from July 1, 1914, to June 30, 1915, Winkoop Hallenbeck, Crawford Co., p. 100.

Warner, D.J. and Ariail, R.B., 1995, Alvan Clark and Sons: Artists in Optics, Willmann-Bell.

Whitesell, P. S., 1998, A Creation of His Own: Tappan's Detroit Observatory, Bentley Historical Library.

# Appendix I. A Little about the Science

During its fifty years of operation, observers at the second campus observatory targeted many objects with many scientific goals in mind. Although this research did not entirely exclude extragalactic objects, stars of various sorts were the main subjects of investigation, a choice partly dictated by the telescope's modest aperture and midwestern location. Below I give a brief introduction to a few reoccurring scientific strands in this research.

*Period Changes:* For eclipsing variable stars, the period is the time interval between two successive primary eclipses. For stars that pulsate, the period is the time needed for a particular pulsation cycle to repeat. In the case of a Cepheid or RR Lyrae star which is pulsating in a single mode, the period equals the time between two successive peaks in brightness. Because some variable stars pulsate simultaneously in more than one mode, they may have more than a single pulsation period. Often the period can be determined far more accurately than any other property of an eclipsing or pulsating star.

The periods of both eclipsing binary stars and pulsating stars can change, but for different reasons. In either case, period changes are of interest because they can tell us about the changing physical properties of the stars or systems in which they occur. Period changes of pulsating and eclipsing stars can be revealed through long term photometric monitoring. Such monitoring has been the subject of several papers employing observations made at the campus observatory. For example, Linnell (1980) used old and new photometry to study the long-term period behavior of the eclipsing binary VW Cephei. He concluded that its changing period was telling us that material was being transferred through space from the more massive to the less massive star within that binary system.

For a pulsating star, the period will change as the star's density changes, in accordance with the pulsation equation: $P \sqrt{\rho} = Q$, where P is the period, $\rho$ is the density of the star, and Q is the pulsation constant. As a pulsating star evolves through the Hertzsprung-Russell (H-R) diagram, its period will thus change inversely to its change in density, provided that Q remains constant. For example, if a star increases in brightness without changing in mass or surface temperature, it must necessarily be growing larger in diameter and its density must therefore be decreasing. Consequently, its period will respond by becoming longer. Stellar evolution theory predicts how quickly RR Lyrae stars and Cepheids should move through the H-R diagram, and thus how quickly their periods should change. Observations of period changes can therefore be used to witness stars evolving and to test theories of how stars should change as they age.

Papers studying the period changes of pulsating stars include, among others, Silbermann and Smith (1995a), who compared observed and theoretical rates of period change for RR Lyrae stars in the globular cluster M15, and Osborn et al. (2019), who studied the long-term period behavior of type II Cepheids in M13.

*Multimode RR Lyrae stars:* As mentioned above, some stars pulsate simultaneously in more than a single pulsation mode. Some of these different pulsations are reasonably well understood but others are more puzzling. RR Lyrae stars have a main pulsation period of about half a day. However, Blazhko effect RR Lyrae stars also show secondary periods much longer than their half-day period. The mechanism behind these long secondary periods, which can take tens of days, is still a matter of controversy. Another type of multimode RR Lyrae star, RRd stars, is better understood. They are believed to be pulsating simultaneously in the fundamental and first overtone radial pulsation modes. However, even for RRd stars, the physics behind the excitation of the dual pulsation modes remains puzzling.

Observations of multimode RR Lyrae stars designed to shed light upon their nature were often carried out at the observatory from the 1990s until the 2010s. Among the resulting papers are LaCluyzé et al. (2004), which deals with changes in the Blazhko effect in XZ Cygni.

*Cataclysmic variable stars:* Cataclysmic variable stars run the gamut from dwarf novae to recurrent novae to classical novae to supernovae. All undergo sudden outbursts in brightness, but the physical processes underlying the outbursts differ and pose a variety of perplexing questions for astronomers. Outbursts of dwarf novae can repeat every few months, while supernova explosions are one-time happenings, in which much of the star (or stars) is exploded into space or locked into a neutron star or black hole. Although cataclysmic variables were occasionally observed during the 1980s, 1990s, and 2000s, photometry of such stars increased during the 2010s, becoming a major research subject. Those observations can test models of different sorts of cataclysmic variables. Among the campus observatory papers dealing with cataclysmic variables are Richmond and Smith (2012) and Patterson et al. (2017a).

*Exoplanets:* Observations related to exoplanets began to be made toward the end of the 2010s. Transits, in which a planet moves in front of a star, slightly dimming its light, are one of the ways by which exoplanets are discovered. However, care must be taken to distinguish between exoplanet transits and eclipses of binary stars. Photometric observations at the second observatory help to verify exoplanet candidates identified in the KELT survey.

# Appendix II: Published Papers

Below I provide a list of publications based in whole or in part on observations with or instrumentation at the MSU campus observatory. I have tried to be as inclusive as possible. For some papers MSU provided the whole of the observations. In other cases, MSU observations form only a small portion of a larger study. The list is divided according to the nature of the publication. Abstracts, conference proceedings, papers in refereed journals, and dissertations are listed separately.

## Abstracts and Notices

Carroll, R.W., Stoeckley, T., Cruddace, R.G., and Hutchings, J., 1979, Optical Observations of Known X-ray Emitting W UMa Stars, Bulletin of the AAS, **11**, 722.

De Lee, N. M., Brueneman, S., Hicks, L., Russell, N., Kinemuchi, K., Pepper, J., Rodriguez, J., Paegert, M., and Smith, H. A., 2017, KELT RR Lyrae Variable Stars Observed by NKU Schneider and Michigan State Observatories, Bulletin of the AAS, **229**, 152.08.

Gay, P. L., Smith, H.A., Braden, J.G., and Silbermann, N.A., 1995, Coupling of Period Change to Pulsation Mode in RR Lyrae Stars, Bulletin of the AAS, **27**, 102.09.

Hill, S. J., and Linnell, A. P., 1974, An Automated Photometric Telescope, Bulletin of the AAS, **6**, 218.

Lacluyzé, A., Smith, H. A., Gill, E. M., Hedden, A., Kinemuchi, K., Rosas, A. M., Pritzl, B. J., Sharpee, B., Robinson, K. W., Baldwin, M., and Samolyk, G., 2002, The Changing Blazhko Effect of XZ Cygni , Radial and Nonradial Pulsations as Probes of Stellar Physics, ASP Conference Proceedings, **259**, 416.

Lin, H. and Kuhn, J. R., 1991, Precision Visible and Infrared Solar Photometry, Bulletin of the AAS, **23**, 23.11.

Linnell, A. P., 1978, An Automated Photometer, Bulletin of the AAS, **10**, 413.

Linnell, A. P., 1978, High Speed Photometry of VW Cephei, Bulletin of the AAS, **10**, 608.

Patterson, J. et al. 2017b, OV Bootis: Forty Nights of World Wide Photometry, Journal of the AAVSO, **45**, 224.

Rabidoux, K., Smith, H. A., Wells, K., Randall, J., Hartley, D., LaCluyze, A., De Lee, N., Ingber, M., Ireland, M., Kinemuchi, K., Pellegrini, E., Purdum, L. E., Pritzl, B. J., Lustig, R., Osborn, W., Lacy, J., Curtis, M., and Smolinski, J., 2005, Light Curves for Type II Cepheids in M3 and M5, Bulletin of the AAS, **37**, 122.07.

Rabidoux, K., Smith, H. A., Randall, J., Wells, K., Taylor, L., Hartley, D., Greenwood, C., Kuehn, C., LaCluyzé, A., De Lee, N., Ingber, M., Ireland, M., Kinemuchi, K., Pellegrini, E., Purdum, L. E., Pritzl, B. J., Lustig, R., Osborn, W., Lacy, J., Curtis, M., and Smolinski, J., 2007, Period Changes and Light Curves of Type II Cepheids in Globular Clusters, Bulletin of the AAS, **39**, 60.13.

Randall, J. M., Rabidoux, K., Smith, H. A.; De Lee, N., Pritzl, B., and Osborn, W., 2007, Period Changes of Type II Cepheids in the Globular Cluster M5, Bulletin of the AAS, **39**, 254.14.

Rivera, F., Ressler, R., Kinemuchi, K., Smith, H. A., 2006, Photometric Observations of RR Lyrae Stars at Red Buttes Observatory, Bulletin of the AAS, **38**, 29.04.

Seki, T., Furuta, T., Tsuchiya, K., Takeishi, M., Urata, T., Balam, D. D., Miller, R. D., Marche, J. D., Maley, P., Morris, C. S., and Marsden, B. G., 1978, Comet Meier (1978f), IAU Circular No. 3227.

Osborn, W., 1977, Basic Observational Data for the Variables in M13. Bulletin of the AAS, **9**, 342.

Silbermann, N. A., 1993, RR Lyrae Variables in the Globular Cluster M15, **25**, 123.01D.

Silbermann, N. A., Harrell, M.J., and Smith, H. A., 1993, RR Lyrae Stars in M14 and M15, Bulletin of the AAS, **25**, 63.04.

Silbermann, N.A., Smith, H. A., and Bolte, M., 1993, CCD Photometry of RR Lyrae Stars in NGC 6388 and M15, in IAU Colloq 139: New Perspectives on Stellar Pulsation and Pulsating Variable Stars, **139**, 338.

Smith, H.A., 1998, Photometry of Multimode RR Lyrae Stars, Bulletin of the AAS, **30**, 67.19.

Smith, H. A., Anderson, M., Osborn, W., Layden, A., Kopacki, G., Pritzl, B., Kelley, A., McBride, K., Alexander, M., Kuehn, C., Kilian, A., King, E., Carbajal, D., Lusting, R., and De Lee, N., 2015, Lightcurves and Period Changes for Type II Cepheids in the Globular Cluster M13. Journal of the American Association of Variable Stars Observers, **43**, 255.

Smith, H. A., Kuhn, J. R., and Curtis, J., 1989, CCD Observations of Variable Stars in Globular Clusters, in The Use of pulsating stars in fundamental problems of astronomy, Proceedings of IAU Colloq. **111**, 285.

Smith, H. A., LaCluyze, A., Gill, E.-M., Hedden, A., Kinemuchi, K., Rosas, A. M., Pritzl, B. J., Sharpee, B., Robinson, K., Baldwin, M., and Samolyk, G., 2001, The Changing Blazhko Effect of XZ Cygni, Bulletin of the AAS, **33**, 46.03.

Smith, H.A., Lee, K.M., Silbermann, N.A., Williams, J.S., and Bolte, M.1992, CCD Photometry of the RR Lyrae Star AH Camelopardalis, Journal of the AAVSO, **21**, 64.

Smith, H. A., Oaster, L. E., Kinemuchi, K., and Kolenberg, K., 2005, Observing Multimode RR Lyrae Stars with the Michigan State University Campus Observatory, Bulletin of the AAS, **37**, 122.02.

Stark, M. A., Wade, R. A., Thorstensen, J. R., Peters, C. S., Sheets, H. A., Smith, H. A., Miller, R. D., and Green, E. M., 2006, A New, Bright, Short-period, Emission Line Binary in Ophiuchus, Bulletin of the AAS, **38**, 09.07.

**Conference Proceedings and Unrefereed Papers**

Bragaglia, A., Clementini, G., di Fabrizio, L., di Tomaso, S., Merighi, R., Tosi, M., Ivans, I., Sneden, C., Wilhelm, R., and Smith, H., 2000, Anomalous Pulsation of Field RR Lyrae Variables: Photometric and Spectroscopic Study of CM Leo, BS Com, and CU Com, IAU Colloq.176: The Impact of Large-Scale Surveys on Pulsating Star Research, ASP Conference Proceedings, **203**, 271.

Kinemuchi, K., 2013, To Pulsate or to Eclipse? Status of KIC 9832227 Variable Star, in 40 Years of Variable Stars: A Celebration of Contributions by Horace A. Smith, arXiv:1310.0149v1.

Linnell, A. P. and Hill, S. J., 1974, An Automated Photometric Telescope, in Instrumentation in astronomy II, Society of Photo-optical Instrumentation Engineers, 63.

Linnell, A. P. and Hill, S. J., 1975, The MSU Computer Assisted Photometric System, in Telescope Automation, Edited by Maureen K. Huguenin and Thomas B. McCord. Published by MIT, 1975, 52.

Linnell, A. P., 1980, The Physical Status of VW Cephei, in Close Binary Stars: Observations and Interpretation, IAU Symp. **88**, 505.

Littlefield, C. et al. 2019, The Rise and Fall of the King: The Correlation between FO Aquarii's Low States and the White Dwarf's Spindown, arXiv: 1904.11505.

Osborn, W., 2013, Man Versus Machine: Eye Estimates in the Age of Digital Imaging, in 40 Years of Variable Stars: A Celebration of Contributions by Horace A. Smith, arXiv:1310.0540.

Patterson, J., de Miguel, E., Barret, D., Brincat, S., Boardman, J., Buczynski, D., Campbell, T., Cejudo, D., Cook, L., Cook, M. J., Collins, D., Cooney, W., Dubois, F., Dvorak, S., Halpern, J. P., Kroes, A. J., Lemay, D., Licchelli, D., Mankel, D., Marshall, M., Novak, R., Oksanen, A., Roberts, G., Seargeant, J., Sears, H., Silcox, A., Slauson, D., Stone, G., Thorstensen, J. R., Ulowetz, J., Vanmunster, T., Wallgren, J., and Wood, M., 2017a, OV Bootis: Forty Nights Of World-Wide Photometry, The Society for Astronomical Sciences 36th Annual Symposium on Telescope Science and AAVSO Spring

2017 Meeting. Published by Society for Astronomical Sciences, **36**, 1.

Smith, H. A., 1999, Long Term Observations of RR Lyrae Stars, in Anni Mirabiles, A Symposium Celebrating the 90th Birthday of Dorrit Hoffleit, 65.

Smith, H.A., 2006, Blazhko Effect and Double-mode RR Lyrae Stars, Memorie della Società Astronomica Italiana, **77**, 492.

Smith, H. A., 2013, Period Changes of Mira Variables, RR Lyrae Stars, and Type II Cepheids, in 40 Years of Variable Stars: A Celebration of Contributions by Horace A. Smith, arXiv:1310.0533.

Zhou, A. Y., Hintz, E. G., Schoonmaker, J. N., Rodríguez, E., Costa, V., Lopez-Gonzalez, M. J., Smith, H. A., Sanders, N., Monninger, G., and Pagel, L., 2017, A Pulsational Time-evolution Study for the Delta Scuti Star AN Lyncis, arXiv:1710.03944.

**Refereed journal papers**

Anderson, A. L., Bestman, W. J., Bieszke, S. A., Blanchard, J., McPherson, M. R., McWilliams, T. R., Mikusko, S. T., Morse, E. C., Philpott, T. N., Sasse, P., Slee, D., Ventimiglia, D. A., and Smith, H. A., 1991, CCD Photometry of the RR Lyrae Star SS Leonis, Journal of the AAVSO, **20**,28.

Bisard, W. and Osborn, W., 1978, Concerning Variable Star 7 in M13, Information Bulletin on Variable Stars, No. 1475, 1.

Clementini, G., Di Tomaso, S., Di Fabrizio, L., Bragaglia, A., Merighi, R., Tosi, M., Carretta, E., Gratton, R. G., Ivans, I. I., Kinard, A., Marconi, M., Smith, H. A., Wilhelm, R., Woodruff, T., and Sneden, C., 2000, CU Comae: A New Field Double-Mode RR Lyrae Variable, the Most Metal-poor Discovered to Date, Astronomical Journal, **120**, 2054.

Di Fabrizio, L., Clementini, G., Marconi, M., Carretta, E., Ivans, I. I., Bragaglia, A., Di Tomaso, S., Merighi, R., Smith, H. A., Sneden, C., and Tosi, M., 2002, Anomalous RR Lyrae stars(?): CM Leonis, Monthly Notices of the Royal Astronomical Socety, **336**, 841.

Jurcsik, J., Sódor, Á., Hurta, Zs., Váradi, M., Szeidl, B., Smith, H. A., Henden, A., Dékány, I., Nagy, I., Posztobányi, K., Szing, A., Vida, K., and Vityi, N., 2008, An Extensive Photometric Study of the Blazhko RR Lyrae star MW Lyr - I. Light-curve Solution, Monthly Notices of the Royal Astronomical Society, **391**, 164.

Jurcsik, J., Sódor, Á., Szeidl, B., Kolláth, Z., Smith, H. A., Hurta, Zs., Váradi, M., Henden, A., Dékány, I., Nagy, I., Posztobányi, K., Szing, A., Vida, K., and Vityi, N., 2009, An Extensive Photometric Study of the Blazhko RR Lyrae Star MW Lyr - II. Changes in the Physical

Parameters, Monthly Notices of the Royal Astronomical Society, **393**, 1553.

Karmakar, P., Smith, H.A., and De Lee, N., 2019, The Long-term Period Changes and Evolution of V1, a W Virginis Star in the the Globular Cluster M12, Journal of the American Association of Variable Stars Observers, **47**, in press.

Kolenberg, K., Smith, H. A., Gazeas, K. D., Elmaslı, A., Breger, M., Guggenberger, E., van Cauteren, P., Lampens, P., Reegen, P., Niarchos, P. G., Albayrak, B., Selam, S. O., Özavcı, I., and Aksu, O., 2006, The Blazhko effect of RR Lyrae in 2003-2004, Astronomy and Astrophysics, **459**, 577.

Korista, K. T., et al. 1995, Steps toward Determination of the Size and Structure of the Broad-line Region in Active Galactic Nuclei. 8: An Intensive HST, IUE, and Ground-based Study of NGC 5548, Astrophysical Journal Supplement, **97**, 285.

Kuhn, J. R., Lin, H., and Loranz, D., 1991, Gain Calibrating Nonuniform Image-array Data using only the Image Data, Publications of the Astronomical Society of the Pacific, **103**, 1097.

LaCluyzé A., Smith, H. A., Clark, A. R., Craven, J. C., Ingber, M. A., Lam, K., Lande, J. L., Neir, M. G.; Prichard, M. N., Sheppard, M. R., and Ziethe, J., 2001,

The Changing Amplitude of the Delta Scuti Star AN Lyn, Information Bulletin on Variable Stars, Nr. 5180, 1.

LaCluyzé, A., Smith, H. A., Gill, E.M., Hedden, A., Kinemuchi, K., Rosas, A. M., Pritzl, B. J., Sharpee, B., Wilkinson, C., Robinson, K. W., Baldwin, M. E., and Samolyk, G., 2004, The Changing Blazhko Effect of XZ Cygni, Astronomical Journal, **127**, 1653.

Le Borgne, J. F., Poretti, E., Klotz, A., Denoux, E., Smith, H. A., Kolenberg, K., Szabó, R., Bryson, S., Audejean, M., Buil, C., Caron, J., Conseil, E., Corp, L., Drillaud, C., de France, T., Graham, K., Hirosawa, K., Klotz, A. N., Kugel, F., Loughney, D., Menzies, K., Rodríguez, M., and Ruscitti, P. M., 2014, Historical Vanishing of the Blazhko Effect of RR Lyr from the GEOS and Kepler Surveys, Monthly Notices of the Royal Astronomical Society, **441**, 1435.

Lee, K., Gay, P., and Smith, H. A., 1996, The Blazhko Effect of the RR Lyrae Star V421 Herculis, Publications of the Astronomical Society of the Pacific, **108**, 659.

Li, K.-L., Kong, A. K. H., Hou, X., Mao, J., Strader, J., Chomiuk, L., and Tremou, E., 2016, Discovery of a Redback Millisecond Pulsar Candidate: 3FGL J0212.1+5320, Astrophysical Journal, **833**, 143.

Lin, H., and Kuhn, J.R., 1989, An Imaging, Tunable Magneto-Optical Filter, Solar Physics, **122**, 365.

Lin, H. and Kuhn, J. R., 1992, Precision IR and Visible Solar Photometry, Solar Physics, **141**, 1.

Linnell, A. P., 1980, VW Cephei – Period and Color Changes, Publications of the Astronomical Society of the Pacific, **92**, 202.

Linnell, A. P., 1981, An Automated System for Photoelectric Photometry. II., Publications of the Astronomical Society of the Pacific, **93**, 661.

Linnell, A. P. 1981, UBVRI Times of Minima of VW Cephei, Information Bulletin of Variable Stars, Nr. 1967, 1.

Linnell, A.P. Hill, S. J., and Brandt, E. F., 1975, An Automated System for Photoelectric Photometry, Publications of the Astronomical Society of the Pacific, **87**, 273.

Linnell, A. P. and Brandt, E. F., 1979, A Crystal-controlled Observatory Clock, Publications of the Astronomical Society of the Pacific, **91**, 731.

Littlefield, C, et al. 2016, Return of the King: Time-series Photometry of FO Aquarii's Initial Recovery from its Unprecedented 2016 Low State, Astrophysical Journal, **833**, 93.

Oaster, L., Smith, H. A., and Kinemuchi, K., 2006, A Double-Mode RR Lyrae Star with a Strong

Fundamental-Mode Component, Publications of the Astronomical Society of the Pacific, **118**, 405.

Osborn, W., Kopacki, G., Smith, H. A., Pritzl, B. J., Kuehn, C., and Anderson, M., 2019, Variable Stars in M13. III. The Cepheid Variables and their Relation to Evolutionary Changes in Metal-poor BL Her Stars, Acta Astronomica, **67**, 131.

Osborn, W., Layden, A., Kopacki, G., Smith, H., Anderson, M., Kelly, A., McBride, K., and Pritzl, B., 2017, Variable Stars in M13. II. The Red Variables and the Globular Cluster Period-Luminosity Relation, Acta Astronomica, **67**, 13.

Purdue, P., Silbermann, N. A.; Gay, P., Smith, H. A., 1995, Double-Mode RR Lyrae Stars in the Globular Cluster M15, Astronomical Journal, **110**, 1712.

Rabidoux, K., Smith, H. A., Pritzl, B. J., Osborn, W., Kuehn, C., Randall, J., Lustig, R., Wells, K., Taylor, L., De Lee, N., Kinemuchi, K., LaCluyzé, A., Hartley, D., Greenwood, C., Ingber, M., Ireland, M., Pellegrini, E., Anderson, M., Purdum, G., Lacy, J., Curtis, M., Smolinski, J., and Danford, S., 2010, Light Curves and Period Changes of Type II Cepheids in the Globular Clusters M3 and M5, Astronomical Journal, **139**, 2300.

Richmond, M. W. and Smith, H.A., 2012, BVRI Photometry of SN 2011fe in M101, Journal of the

American Association of Variable Star Observers, **40**, 872.

Silbermann, N.A. and Smith, H.A., 1995a, Period changes of RR Lyrae stars in the globular cluster M15, Astronomical Journal, **109**, 1119.

Silbermann, N. A. and Smith, H. A., 1995b, The RR Lyrae Variable Stars in the Globular Cluster M15, Astronomical Journal, **110**, 704.

Smith, H. A., 1993, CCD Photometry of Variable Stars in a Sophomore Level Astronomy Laboratory, Journal of the AAVSO, **22**, 35.

Smith, H. A., Barnett, M., Silbermann, N. A., and Gay, P., 1999, The Blazhko Effect of AR Herculis, Publications of the Astronomical Society of the Pacific, **118**, 527.

Smith, H. A., Church, J. A., Fournier, J., Lisle, J., Gay, P., Kolenberg, K., Carney, B. W., Dick, I., Peterson, R.C., and Hakes, B., 2003, The Blazhko Effect of RR Lyrae in 1996, Publications of the Astronomical Society of the Pacific, **115**, 43.

Smith, H.A., Matthews, J.M., Lee, K. M., Williams, J., Silbermann, N. A., and Bolte, M., 1994, AH Cam: A Metal-rich RR Lyrae star with the Shortest Known Blazhko Period, Astronomical Journal, **107**, 679

Stark, M. A., Hamlin, M. T., Anderson, J. R., Berryman, F. V., Disbro, M. A., Dyni, C., Gates, T. A., McKinney, J. A., Robinson, K. W., Snavley, S. T., Stoltz, T. J., and Smith, H. A., 2000, The Light Curve of the Eclipsing Binary Star XX Leo, Journal of the AAVSO, **28**, 25.

Stark, M. A., Wade, R. A., Thorstensen, J. R., Peters, C. S., Smith, H. A., Miller, R. D., and Green, E. M., 2008, A New, Bright, Short-Period, Emission Line Binary in Ophiuchus, Astronomical Journal, **135**, 991.

Wehlau, A. and Bohlender, D., 1982, An Investigation of Period Changes in Cluster BL Herculis Stars, Astronomical Journal, **87**, 780.

## MS Theses

Chuang, Steve Shih-Hsein, 1982, Analysis of the Occultation of SAO 93957 (Central Michigan University).

Spalding, Roger D., 1979, BV Photographic Study of the Galactic Cluster Stock 1 (Central Michigan University).

## PhD Dissertations

Lin, Haosheng, 1992, Precision Infrared and Visible Solar Photometry: A Photometric Study of the Solar Irradiance Variation (Michigan State University).

Silbermann, Nancy Ann, 1994, The RR Lyrae Variable Stars in the Globular Cluster M15 (Michigan State University).

# Index

www.ingramcontent.com/pod-product-compliance
Lightning Source LLC
Chambersburg PA
CBHW071231210326
41597CB00016B/2007